The theory of relativity is the confession of relativists that they do not know what relative speed is.

Relativity is a serious crime (fraudulence) against humanity, Nature, and the Creator of universe, if any. Relativists have wasted or stolen billions of tax dollars only to fool humanity and hamper the advancement of physical science for more than one hundred years (1905~). Yet relativists adamantly refuse to check, let alone acknowledge, the fallacy of relativity. Normal academic activities cannot fix this problem. The only solution left is to bring relativity to justice. This book is an indictment against Harvard physicists and all other relativists.

What the lawyers and judges should do is simple; that is they check whether relative speed is observer-*free* or *observer-dependent* through a simple experiment as shown in **Fig. C**, and the truth will out in seconds.

The ***Preface*** and ***Summary*** of this book are for those who would like to check the fallacy of relativity with the least hassle with math. But the highlight of this book is the math in the main body. The author recommends all readers that they review the preface and summary before they check the main body.

Unfortunately, however, due to the shortage of budget and time, the author could not have his manuscript thoroughly proofread. Yet the author thinks that this book is enough to disprove relativity thoroughly and completely.

–Author / August, 2014

Harvard Physicists

Confuse Relative Speed with Proper Speed

Relative Speed vs. Proper Speed

Both space and time are amorphous (formless) entities. The idea that space and time form a four-dimensional unity and that this unity is contracted, expanded, curved, etc. in order to make the speed of light always the same (c) for all human observers and non-human objects, whose motions are truly arbitrary and unpredictable, is a manifold delusion; and this delusion came from not distinguishing relative speed from proper speed.

Byoung Ahn

**Harvard Physicists
Confuse Relative Speed with Proper Speed**

Relative Speed vs. Proper Speed
By Byoung Ahn

Copyright © 2014 by Byoung Ahn.
All right reserved.
Printed by CreateSpace, An Amazon.com Company, USA.
Available from Amazon.com and other retail outlets.
ISBN-13: 978-1500685799
ISBN-10: 1500685798

About the author
- o Graduated from Seoul National University (South Korea)
  (Bachelor and master's degrees in biology)
- o Graduated from Southern Connecticut State University (USA)
  (bachelor's degree in art)

To my father and mother.

Contents

Preface --- 11
Summary --- 19

Part I Special Relativity

1. Distance, Speed, and Velocity ---------------------------------- 53
2. Reference Body and Coordinate System -------------------------- 63
3. Kinds of Dimensional Situations ------------------------------- 65
4. The Case in Which an Observer Sits Off the Line of Motion of a Uniformly-Moving Object --- 69
5. r-factor -- 77
6. The Secrets of Fixed Stars ------------------------------------- 81
7. Einstein Mistook Proper Velocity for Relative Velocity --------- 87
8. Cosine Effect -- 90
9. The Case in Which Two Lines of Motion Meet -------------------- 101
10. The Case of Parallel Motion ---------------------------------- 105
11. The Case of Accelerated Motion ------------------------------- 109
12. The Case of Circular Motion ---------------------------------- 112
13. The Case of Rotation Motion ---------------------------------- 115
14. GPS and Relativity --- 119
15. Relativity of Simultaneity ----------------------------------- 122
16. Einstein's Concept of Inertia Is Incorrect ------------------- 131
17. The Motion of Photons Is Different from That of Ordinary Objects -- 136
18. Are Space and Time Relative or Absolute? -------------------- 138
19. The Gyroscope and Absolute Space ----------------------------- 148
20. The Mistakes of Galileo -------------------------------------- 152
21. Relative to What Is Light Speed Constant? ------------------- 154

22. How to Measure Absolute Motion/Speed -------------------- 161
23. Relativistic Addition of Speeds ----------------------------- 163
24. The Lorentz Transformation (LT) --------------------------- 173
25. The Light-year and Relativity ------------------------------- 179
26. Does the Universe Expand at Faster-than-light Speed? ----------- 183
27. The Michelson-Morley Experiment ------------------------- 187
28. Time Dilation, Length Contraction, and Mass Increase Cannot Be Measured -- 195
29. Twin Paradox --- 199
30. Muon Paradox -- 203
31. Clock Paradox -- 210
32. $E = mc^2$ --- 219
33. The Law of Conservation of Matter and Energy ------------- 226

Part II General Relativity

34. "Relativity of Acceleration" Is Wrong ---------------------- 231
35. Inertial Mass is Not the Same as Gravitational Mass -------- 233
36. Eight Characteristics That Distinguish Gravity from Inertial Force (The Principle of Equivalence Is Wrong) ---------------------- 235
37. Light Does Not Deflect in the Gravitational Field ---------- 238
38. Is Special Relativity Incompatible with General Relativity? ---- 243
39. The Fallacy of the Chest Thought Experiment -------------- 247
40. The Fallacy of the Disc Thought Experiment --------------- 253
41. Gravity vs. the Geometry of Space-time -------------------- 258
42. Eddington's Solar Eclipse Observation --------------------- 260
43. Unified Field Theory -------------------------------------- 276

Appendix-1 Six Exercises of Finding Relative Speed ----------- 285
Appendix-2 Four Laws of Relative Speed ---------------------- 304
Appendix-3 Einstein—His Religion, Philosophy, and Morality -305
Epilogue --- 328
References -- 329
Index --- 336

Preface

Speed is the rate of change in position of an object with respect to a specific reference body. Speed is *scalar* that has *no direction* whereas velocity is *vector* that has a direction. That is, velocity is *speed* plus (+) *direction*. For example, 60km/h is speed; 60km/h, south, is velocity.

What is important in relativity is the *absolute magnitude* (positive number) of the *rate* of change in distance between two specific objects—an observer and a target object, for example; whether the rate of change in distance increases (+) or decreases (−) does not matter in relativity.

Relative speed is *reciprocal*. That is, the speed of object *A* relative to object *B* is the same as the speed of object *B* relative to object *A*. But *relative velocity* (speed + direction) is not reciprocal because the relative *direction* of one party is the opposite of the relative direction of the counterpart object. That is, reciprocity holds only in *relative speed* and not in *relative velocity*. Therefore, it is reasonable and convenient to use the term *speed* than *velocity* in relativity. For this reason the author uses mainly *speed* in this book unless the *direction* of motion is at stake.

The terms and concepts of *relative speed* and *proper speed* in this book (see the illustration on the back cover) are authors if not indicated otherwise. Relativists use their own terms *velocity* and *proper velocity*, both of which are totally different in concept from that of the authors. The problem with relativists' terms *velocity* and *proper velocity* will be discussed later.

Another problem with relativity is that relativists deal with speed *in terms of 1-D (one-dimensional) situation* even when the given situation is actually 2-D or 3-D.

The 1-D situation means the case in which two involved objects are situated or moving on the same straight line. For example, in **Fig. C** (see the front cover), the car and Einstein are in a 1-D situation. In a 1-D situation, the relative speed between two objects is simply *the difference* between the proper speed of one party (v_1) and that of the other (v_2). For example, the proper speed of the car in **Fig. C** is v and that of Einstein is zero (0); therefore, the relative speed between the car and Einstein is $v \sim 0 = v$. (The state of being stationary of an object in a given inertial system is a kind of *proper motion*. In this case the value of the proper speed is zero <0>). However, when it comes to 2-D or 3-D situations, the relative speed between two involved object is not simply $v_1 \sim v_2$; we should use calculus or trigonometry or radar in order to find the relative speed between the two.

A 2-D situation means the case in which two involved objects are *not* situated/moving on the same straight line yet both are on the same *plane*. For example, in **Fig. C,** the car and Galileo are in a 2-D situation.

A 3-D situation is the case in which two involved objects move independently of each other but the line of motion of one party and that of the other are not in the same plane. (See the illustration of various cases of 1-D, 2-D, and 3-D situations in **Chapter 3** and **Appendix-1**).

The term *line of motion* means a straight line along which an object moves (author's term/definition). If an object is in curvilinear motion, the tangent line at a given point on the curvilinear path of the object is the line of motion of the object at the given moment.

By the way, the relativistic formula of speed addition/subtraction ($\dfrac{v_1 \pm v_2}{1 + \dfrac{v_1 \times (\pm v_2)}{c^2}}$) is applicable only to 1-D situations; this formula is *not* applicable to 2-D or 3-D situations. Therefore, relativists cannot find the relative speed between the car and Galileo in **Fig. C** by using the speed addition formula (because the car and Galileo are in a 2-D situation).

Relativists' concept of speed/velocity is *incorrigibly* vague and incorrect; this problem is found in all literature that deals with relativity. *The twin paradox* is a typical example. The twin paradox explains as follows:

"If a twin brother takes a space trip on a spacecraft to a star at the uniform speed v relative to his twin sister on the earth, the distance the spacecraft covers is **contracted** by the factor of $\sqrt{1 - v^2/c^2}$ (where v is the relative speed between the two twins and c is the speed of light) when viewed from the twin sister on the earth. The same phenomenon of *distance contraction* occurs when the spacecraft returns home." These stories are a part of the twin paradox.

> However, relativists say exactly the opposite when they explain t*he muon paradox*: Relativists say that the distance a cosmic particle muon covers is *elongated* (not *contracted*!) by the factor of $\dfrac{1}{\sqrt{1-v^2/c^2}}$ (note that the denominator is always smaller than 1) when viewed from a stationary observer on earth, toward which the muon darts from high altitude at the speed of v. This is *distance elongation* and not *distance contraction*. Here we see a major contradiction between the twin paradox and the muon paradox (more on this in **Chapters 29 and 30**).
>
> By the way, according to the isotropy of light speed, $c + v = c$ and $c - v = c$. Then,
>
> $$\sqrt{1-v^2/c^2} = \sqrt{\dfrac{(c+v)(c-v)}{c^2}} = \sqrt{\dfrac{c \times c}{c^2}} = 1, \text{ and}$$
>
> $$\dfrac{1}{\sqrt{1-v^2/c^2}} = 1.$$
>
> The above results show that the Lorentz transformation (LT) is self-contradictory and that distance contraction/elongation is not real (more on the LT in **Chapter 24**).

As for now, our concern is neither distance contradiction nor distance elongation. Our concern now is the fact that relativists' concept of relative speed is helplessly vague and incorrect. Let us come back to the story of the twin paradox:

Relativists *assume, without clear thinking*, that the two twins maintain a uniform relative speed v throughout the space trip experiment. This is a huge and fundamental mistake. If the twins are to maintain a uniform relative speed, say v, the two twins must remain on the same straight line throughout the space trip. In other words, the two twins need to maintain a 1-D situation throughout the space trip. But that is impossible in reality; as we can see in **Fig. P-1** (see next page), the earth, on which the twin sister rests, rotates and revolves around the sun, and the solar system and the galaxy are also in complicated motion of their own in space.

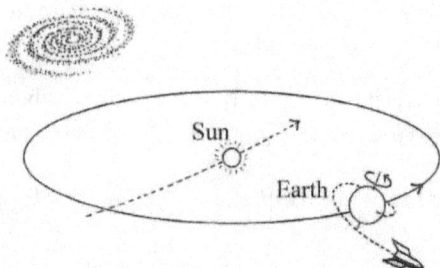

Fig. P-1 The earth is in complicated motion in space.

Obviously, the two twins cannot remain in a 1-D situation. That is, it is impossible for the two twins to maintain a uniform relative speed. This is so even if we assume that the spacecraft maintains uniform linear motion (with respect to *absolute space*) with the help of **gyroscopes** (more on absolute space and the gyroscope in **Chapter 19**). Besides, there are many other fundamental problems with the twin paradox (see **Chapter 29**).

The custom of considering relative speed in terms of 1-D situation and the custom of misidentifying proper speed (observer-free speed) with relative speed (observer-dependent speed) came from Einstein. Einstein thought that all motion/velocity is relative (Einstein used the term *velocity*); he did not distinguish the difference between proper speed from relative speed; he took (mistook!) proper speed for relative speed; he considered speed only in terms of 1-D situation even when the given situation was actually 2-D or 3-D situation (more on these in **Chapter 7**).

Relativists are, as was Einstein, attached to the idea that relative speed exists/occurs only in a 1-D situation. Relative speed in a 2-D or 3-D situation is *unthinkable* or *unimaginable* for relativists. That is, relativists' concept of speed is **one-dimensional**.

Relativists may argue that relativity deals with *velocity* in *four-dimensional* space-time. Whatever the four-dimensional space-time may be, relativists consider the speed of an object based on the *assumption* that both the object and an observer are situated *on the same straight line*. As long as two objects are situated or moving on the same straight line, **regardless of** whether the straight line is curved or contracted or twisted or whatever, the two objects are in a 1-D situation (see **Fig. P-2**).

Preface

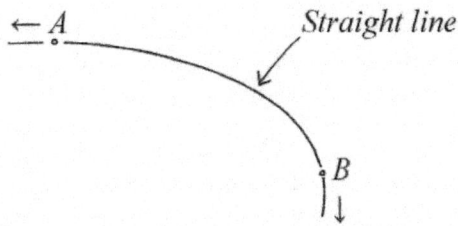

Fig. P-2 If two objects are situated or moving on the same straight line, regardless of whether the straight line curved or not, it is a 1-D situation.

The idea that a straight line is curved, elongated, or contracted in four-dimensional space-time is another delusion to debunk (see **Chapter 28**).

From the days of Galileo (1564-1642) to date, not to mention the days prior to Galileo, all scientists including Galileo, Newton, Einstein, Harvard physicists, etc. have thought of speed only in terms of 1-D situation and they have not distinguished proper speed (observer-free speed) from relative speed. The author requests that all physicists (both relativists and dissenters) make clear whether their term *speed* is proper speed or relative speed and whether their concept of relative speed is of 1-D or 2-D or 3-D when they make any serious statement or argument concerning relativity.

Modern relativists use their own terms *velocity* and *proper velocity*, each of which is of totally different in concept from that of the authors. Wikipedia explains relativists' terms *velocity* and *proper velocity* in the article "Proper velocity" as follows:

> In relativity, proper velocity, also known as celerity, is an alternative to velocity for measuring motion. Whereas velocity relative to an observer is distance per unit time where both distance and time are measured by the observer, proper velocity relative to an observer divides observer-measured distance by the time elapsed on the clocks of the traveling object. Proper velocity equals velocity at low speed. Proper velocity at high speed, moreover, retains many of the properties that velocity loses in relativity compared with Newtonian theory.
>
> -------------------------------------
> (Source:http://en.wikipedia.org/wiki/Proper_velocity,"Proper Velocity.")

The above statement sounds like the language of aliens for readers who are not familiar with relativity. The author explains the alien language above as follows for readers:

- The relativists' term velocity, which means relative speed, is similar to the author's term proper speed, which is observer-free.
- The relativists' term proper velocity is a kind of relative speed based on the assumption that distance contraction and time dilation really occur.
- Both the relativistic terms velocity and proper velocity are one-dimensional (1-D) in concept; both terms are defined based on the assumption that an observer and a target object are situated on the same straight line with the observer being stationary on the straight line.
- Consequently, the act of explaining or supporting relativity using the relativistic terms velocity and proper velocity, the latter of which is based on the assumption that distance contraction and time dilation really occur, is circular reasoning.

According to special relativity, relativistic changes (distance contraction, time dilation, etc.) occur only in one-direction—the direction of motion of an object. The Lorentz factor ($\sqrt{1-v^2/c^2}$) and the formula of speed addition are applicable only to 1-D situations. The Lorentz transformation (LT) discusses the observer-free speed (=observer-obsolete speed or proper speed) of an object in two different inertial environments/systems. In other words, the LT deals with proper speed (see **Chapter 24**). Relativists' concept of light speed is one-dimensional because relativists assume that a stationary observer is aligned with the line of motion of a photon. These stories show that special relativity is a ***one-dimensional ideology***. By the way, general relativity is the extension of special relativity (see **Chapters 39-40**.)

Relativistic changes do not occur; even if they did, there were no way for scientists to check or measure such effects with a ruler or watch or scale because these measuring devices are also subject to undergo the same rate of relativistic changes. Relativistic effects are like a mirage or a rainbow that disappears if pursuers come close to.

The author summarizes the highlights of this book as follows:

Preface

- The discovery of the difference between proper speed, which is observer-free, and relative speed, which is observer-dependent.
- The discovery that relativity deals with speed in terms of 1-D situation; relativity does not deal with relative speed in 2-D or 3-D situations.
- The clarification of the difference between actual motion/speed and virtual motion/speed (see **Chapter 1**).
- The discoveries of a series of equations that helps find relative speed in various kinds of dimensional situations (1-D, 2-D, and 3-D) (see **Chapters 4-11** and **Appendix-1**).
- The logical argument that space and time are absolute entities and that space is absolute without ether (see **Chapters 18-19**).
- The logical argument that the proper speed of light is constant with respect to absolute space and that the relative speed of light with respect to observers is not constant (**Chapters 17** and **21**).
- The discovery of *the four laws of relative speed*. The author maintains that physicists cannot say they understand the concept of relative speed until they fully understand the four laws of relative speed (see **Chapters 5** and **6** and **Appendix-2**).
- The discovery of the eight different characteristics that clearly distinguish inertial force, which Einstein misidentified with gravity, from gravity, which comes from the body of matter (see **Chapter 36**).

The theory of relativity came from not distinguishing relative speed from proper speed. Relativity is the confession of relativists that they do not know what relative speed is.

The relativistic argument that four-dimensional space-time continuum contracts, expands, or curves (in order to make the speed of light always the same for all observers) is non-science. The Big Bang, black whole, dark matter (unknown matter), dark energy (unverifiable energy), worm hole, time travel, multiverse, etc. are fairy tales that do not exist in reality. Relativity is not simply a fairy tale or delusion. Though not intentional, the belief in relativity has grown to be a colossal organized crime against humanity, Nature, and the Creator of universe, if any. The belief in relativity is a kind of cult or religion. If a cult grasps secular power, it becomes a religion. The power of a cult comes from not knowing what the core of the cult is.

Colleges, institutes, mainstream media, and even all lay people are brainwashed with relativity or the *Einstein cult*. Since relativity is the source of power, money, and self-esteem for relativists, relativists never acknowledge the fallacy of relativity. The author has confirmed for the past 13 years (2001-2014) that relativity cannot be corrected by normal academic activities. The only solution left is to bring relativity to justice. This book is an indictment against Harvard physicists and all other relativists.

The lawyers and judges do not need to check all the details of the abstruse lies of relativity. All the judges need is to check whether **relative speed is observer-free or observer-dependent through** a simple demonstrative experiment as shown in **Fig. C**. The result will out in seconds: if the speed of the car with respect to Galileo (or any observer who is off the line of motion of an object) turns out to be constant, relativity is true; if not, relativity is false.

The **Summary**, which follows immediately, is for those who would like to review the author's arguments with the least hassle with math. The main body of this book is written in the first person.

–Author / August, 2014 ♦

Summary

1. Relativity does not deal with relative speed.

The theory of relativity, as its title means, should deal with relative speed. However, as we have seen in the cover illustration and the preface, relativity does not deal with relative speed; it deals with proper speed, which is observer-free.

2. Relativity deals with speed *in terms of* 1-D situation.

Physicists—not only relativists but also dissenters!—deal with speed only in in terms of 1-D situation even when the given situation is actually 2-D or 3-D. Physicists customarily think that relative speed exists/occurs only in 1-D situations and never in 2-D or 3-D situations.

Here is a remedy for those who have trouble understanding that relative motion/speed occurs not only in 1-D situations but also in 2-D and 3-D situations: Imagine that everything in the scene in the diagram below is *removed* or becomes invisible except Galileo and the car.

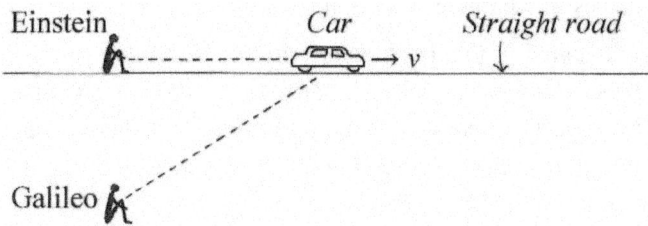

Fig. S-1 Imagine that everything except Galileo and the car is removed or becomes invisible. Yet Galileo and the car are still in a 2-D situation.

Now the situation is somewhat like Galileo and the car are in space. (Readers can replace the car with a spacecraft if they are uncomfortable with a car moving in space.) Yet the car and Galileo are still in a 2-D situation because Galileo is *off* the line of motion of the car. In this situation, the *rate of change* in distance between the car and Galileo is *not uniform* (v) but **changes continuously** as the car moves. This phenomenon is consistently confirmed by radar (speed detector or speed gun), calculus, and trigonometry (cosine effect) (see **Chapters 4 - 8**).

Not a single physicist among those whom the author has contacted so far (2001~) understood that the relative speed between the car and Galileo is less than v and changes continuously with the passage of time.

A member of the NPA (Natural Philosophy Alliance), whom the author knew on the Internet, argued that the direction of the car (in **Fig. S-1**) with respect to Galileo changes continuously as the car moves the observer (Galileo) has to turn his head continuously in order to observe the car, so that the relative speed between Galileo and the car is not determined. This means that the car is not in linear translator motion with respect to Galileo.

Theoretically, we can replace both the car and Galileo with two *points* or very small objects respectively. Yet there still occurs relative speed (*instantaneous relative speed*) between the two regardless of whether either or both of them are in linear translator motion or curvilinear rotatory motion independently of each other. Note that radar (speed gun) can read the speed (*instantaneous relative speed*) of a baseball (or bullet or anything) with respect to the speed gun regardless of whether the baseball spins in whatever direction at whatever RPM (round per minute) and regardless of whether the trajectory of the baseball curves in whatever direction.

Einstein stated in this book *Relativity* (15th ed.) that relative velocity (in fact *proper sped* in the author's term) is available only when a target object is in uniform linear *translatory motion*. (Translatory motion is the opposite of *rotatory motion*. See **Chapters 6** and **7**). [1] [2]. But such a restriction is meaningless or unnecessary.

Radar (speed detector or speed gun) is a very convenient machine; radar is designed to measure the speed of a (any) target object relative to the radar (or the observer who holds the radar). Radar works regardless of whether the target object or the radar itself or both of them are in uniform translator motion or non-uniform rotatory motion or any other kind of complicated motion. Radar works regardless whether the target object and the radar are in a 1-D or 2-D, or 3-D situation. When it comes

Summary

to measuring the relative speed (instantaneous relative speed) of ordinary objects, radar is an almighty machine.

Radar was invented by ***practical scientists***, not by theoretical or armchair scientists. Relativity is an armchair discussion, and relativists are armchair scholars. I have not met a single relativist who understands what radar has to do with relative speed. Relative speed is what radar reads! Radar can read proper speed, too; if an object is in uniform motion in a given inertial system and a radar sits in the line of motion of the object, the speed appears on the radar is the proper speed of the object. Radar reads the *instantaneous relative speed*, not average relative speed.

By the way, what is important in relativity is ***instantaneous relative speed*** and not ***average relative speed***. If relativistic changes (length contraction, time dilation, etc.) really occur, such changes must occur *instantaneously* in accordance with ***instantaneous relative speed***.

Proper speed is instantaneous and linear (one-dimensional) in concept because it occurs on a straight line. Instantaneous relative speed between any two objects is also linear in concept. The relative motion of an object with respect to an observer can be uniform or non-uniform or linear or non-linear. However, the *instantaneous* relative motion/speed between the two is linear in concept even when the object (or the observer) is in non-linear motion.

3. Reference body, environment, and coordinate system

The motion or speed of an object is defined with respect to a certain *reference body*. The author classifies reference bodies into two categories—***environmental reference body*** and ***point reference body***.

Environmental reference body (simply *environment*) is an inertial environment such as a line (straight line), plain, or body of space relative to which the ***actual*** or ***proper*** motion/speed of an object is defined (more on actual motion/speed a moment later).

Point reference body is an inertial object or observer that can be represented or treated as a ***point*** from which the relative motion or speed of a target object is defined.

The ***proper speed*** of an object is defined in (or with respect to) an inertial environmental reference body such as a straight line (1-D environmental reference body), plain (2-D environmental reference body), or a body of space (3-D environmental reference body). But the *relative speed* of an object is defined with respect to a specific *point reference body* such as an observer or counterpart object or *observation point*.

We *cannot* define the relative speed of an object with respect to *an environmental reference body* such as a straight line or plane or body of space because these (a straight line, plain, and body of space) are not represented by a point.

We need to note that the concept of a *coordinate system* should be an environmental reference bodies. In other words, a point reference body (such as an observer or observation point) cannot be a coordinate system because a point does not have coordinates whatsoever. Einstein thought that a point reference body (= an observer or object) can be a coordinate system. Only environmental reference bodies (such as a straight line, plain, and a body of space) have coordinates.

Einstein thought that we can put three perpendicular planes to any point or object (=point reference body) so that we can define the relative velocity (speed + direction) of a target object with respect to the point reference body. However, if three perpendicular planes are attached to a point reference body, it is already a three-dimensional environmental reference body. If there is a point reference body and a target object, we can define only the *relative speed* (not *relative velocity*) of the target object with respect to the point reference body; we cannot define the *direction of motion* of the target object with respect to a point reference body. The direction of motion is defined only in an coordinate system.

Einstein did distinguish environmental reference bodies from the point reference bodies. This is the reason his theory (relativity) does not discern proper speed from relative speed.

The author does not recognize 4-D (3-D space + time) or higher dimensional coordinate systems; these are the imagination of relativists who do not (cannot) have clear understanding in coordinate system.

We can discuss the geometry or curvature of a curved surface using non-Euclidian geometry. However, the idea that space and time, both of which are amorphous entities, form a four-dimensional unity and this unity is curved like the surface of a saddle or funnel is nonsense. The fact that non-Euclidian geometry is useful and true (in the study of curved surface) does not necessarily support the four-dimensionality of relativity.

Summary

4. Actual motion/speed vs. virtual motion/speed

Actual motion is *the **actual change*** in position of an object in a given inertial environment/system, and actual speed is the ***rate of actual motion*** (author's definition). Virtual motion is *any motion that is **not** actual* in a given system. The ability to distinguish actual motion/speed from virtual motion/speed is very important in understanding the fallacy of the theory of relativity. Let us see why:

In **Fig. S-1** (see page 19), the car is in actual motion at the actual speed v in the given inertial environmental reference body (straight road or plane or the body of space). But Einstein and Galileo are stationary in the given environment. In this case, the actual speed of Einstein and that of Galileo are zero (0) respectively. The state of being stationary of an object in a given system is considered a kind of actual motion; in this case the actual speed is zero (0) in value.

Proper motion is a kind of actual motion whereas relative motion is either actual or virtual. For example, the relative motion/speed (v) of the car (in **Fig. S-1**) with respect to Einstein is *actual one* because the car is in actual motion in the given environment. However, the relative motion/speed (v) of Einstein with respect to the car is *virtual one* because Einstein is not in actual motion in the given environment.

Galileo (in **Fig. S-1**) and the car are in relative motion with each other at *ever-changing* relative speed. In this case, the relative motion/speed of Galileo with respect to the car is virtual one because Galileo is not actual motion in the given system. Likewise, the relative motion/speed of the car with respect to Galileo is also a virtual because the car is not actually in such relative motion/speed (ever-changing speed) in the given system.

Proper motion/speed is a kind of actual motion/speed. The difference between proper motion and actual motion is that the former is uniform and linear whereas the latter may be uniform, non-uniform, linear, or curvilinear. That is, actual motion/speed is not necessarily linear and uniform.

As is proper speed, actual speed is observer-free or observer-independent. That is, actual speed is not affected by the motion or position of observers. For example, the speed appears on the speedometer of a car is the actual speed of the car regardless of whether the car is in uniform motion, non-uniform motion, linear motion, or curvilinear motion. No observers outside the car can affect the speed that appears on the speedometer of the car.

What does the difference between actual motion and virtual motion signify? It signifies a lot:

The relativistic idea of length contraction came from the actual motion of an object in **absolute space** allegedly filled with *material ether*. In Einstein's time many scientists believed that space was filled with material ether and space is absolute. They thought that if an object moves (this is the actual motion in absolute space) in the sea of ether, the length of the object would be reduced or compressed a bit in the direction of the *actual motion* of the object due to the pressure or resistance of the *material ether*.

So length contraction (and time dilation), if it were true, would occur only when an object moves actually in absolute space filled with ether. In other words, length contraction and time dilation would not occur if an object is stationary with respect to the ethereal absolute space no matter at what *relative speed* the object moves (this is virtual motion) with respect to a third party in actual motion. The idea that length contraction would occur in actual motion makes sense as long as space were filled with material ether. But ether was controversial material even in Einstein's time because scientists failed to detect such material.

Lorentz (1853-1928), who believed in ether and absolute space, was explained that length contraction and time dilation occur at the rate of $\sqrt{1 - v^2/c^2}$, which is referred to as the *Lorentz factor*. Lorentz formulated this factor and the formula of the addition of speeds in order to explain the isotropy of light speed. However, like his contemporaries, Lorentz did not distinguish actual motion from virtual motion.

Einstein borrowed Lorentz's ideas and math, which were based on ether theory, when he formulated the special theory of relativity (1905). General relativity (1915) is the extension of special relativity (see **Chapters 39-40**). Therefore, Einstein's relativity is a kind of ether theory or the extension of ether theory. But Einstein discarded the controversial ether because he found that his math in his paper did not need ether. The math does not discern actual motion from virtual motion!

The existence of ether had been questioned even before the theory of relativity came. Relativists are proud of the feat of Einstein that

Summary

Einstein put an end to the controversy over the existence of ether and the absoluteness of space and time. The **Michelson and Morley Experiment** (MME, 1887) was interpreted by relativists as the evidence of the isotropy of light speed, the relativeness of space and time, and the non-existence of ether. But the MME had nothing to do with the isotropy of light speed and the relativeness of space. Michelson himself did not accept the relativistic interpretation of his experiment (see **Chapter 27**).

At least ether-believers had the rationale of the length contraction and time dilation effects because they believed in ethereal pressure//resistance. By discarding ether, however, Einstein and his followers have lost such rationale. Relativists believe in the math of relativistic changes and discard the cause (ether and actual motion in the sea of ether). Relativists invented all kinds of thought (theoretical) experiments to prove relativistic changes. Relativists drew the conclusion first and then invented theoretical experiments to support the pre-drawn conclusion. This is not a proper method of science.

5. Length contraction occurs only in uniform linear motion?

According to relativists, special relativistic changes such as length contraction and time dilation occur only in uniform linear motion. But the author thinks that if relativity is true, relativistic changes should occur not only in uniform linear motion but also in any kind of motion including non-uniform motion, curvilinear motion, accelerated motion, etc. Let us see why so.

In the diagram below, (a) is the case of *uniform* motion, (b) is of *intermittent* acceleration, and (c) is of *continuous* acceleration.

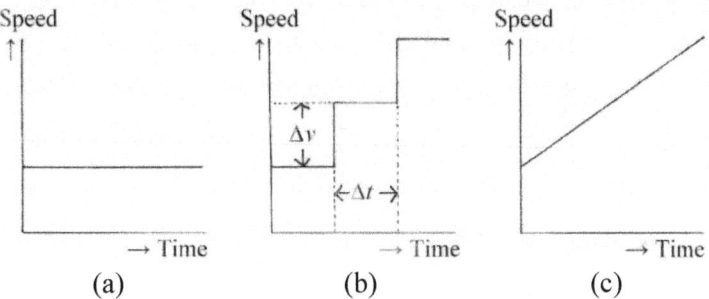

Fig. S-2 Uniform motion (a), intermittent acceleration (b), and continuous acceleration (c)

In graph (b) in **Fig. S-2**, Δt and Δv are not necessarily uniform (constant) respectively. In (c), the speed graph is not necessarily straight; if acceleration is uniform, the graph of (c) is straight; if acceleration is not uniform, the graph is not straight. Now, if relativity is true, the following must be true:

- In graph (a), the rate of relativistic changes is uniform (constant).
- In graph (b), the rate of relativistic changes increases in step-like pattern. Yet during each Δt, the rate of relativistic changes is constant (uniform).

What about case of (c)? The author thinks that the *rate* of relativistic changes increases continuously. The reason or logic is as follows:

If Δv in graph (b) is reduced to zero (0) ($\lim_{\Delta v \to 0}$), graph (b) becomes graph (a). If Δt in (b) is reduced to zero ($\lim_{\Delta t \to 0}$), graph (b) becomes graph (c). Considering that relativistic changes occur at each and every instantaneous relative speed, it is a logical conclusion that relativistic changes occur not only in the cases of (a) and (b) but also in (c).

There is no difference between uniform motion and non-uniform in the point that both cases consist of *instantaneous speeds*. If relativity is true, relativistic changes must occur *instantaneously* in accordance with *instantaneous relative speed* regardless of whether the instantaneous relative speed is constant (this is the case of uniform motion) or changes (this is the case of non-uniform motion).

We have already seen that relative speed occurs not only in 1-D situations but also in 2-D or 3-D situations. Logically speaking, then, special relativistic changes should occur not only in uniform linear motion but also in all kind of motion and in any kind of situation.

However, the author does not mean that relativistic changes (length contraction, time dilation, mass increase, etc.) really occur in any kind of motion and in any kind of situation. Relativistic changes do not occur at all.

6. Einstein returned to ether

Einstein had formulated special relativity (1905) based on the negation of ether and the absoluteness of space and time. But later, in 1920, in his public address at the Leiden University, he changed his mind and recognized *"an ether"* and the absoluteness of space and time. [3] Einstein observes in his address as follows:

> According to the general theory of relativity, space is endowed with physical qualities: in this sense, there exists an ether. Space without ether is unthinkable; for in such space there not only would be no propagation of light, but also no possibility of existence for standards of space and time (measuring-rods and clocks), nor therefore any space-time intervals in the physical sense. But this ether may not be thought of as endowed with the qualities of ponderable media, as consisting of parts which may be tracked through time.

Einstein's statement above means that he acknowledges the absoluteness of space and time. But Einstein's ether is still as elusive as the ether of the previous version. Einstein describes his ether as a medium through which light propagates. He maintains the idea that light needs medium to propagate through in space. **Newton's emission theory** or **corpuscular theory** of light does not need material medium. [4]

Light is known to have properties of both particle and wave. The author thinks that light can make the most of its particle property and propagate through space without medium (ether). The author also thinks that though space looks empty it has absolute capacity to hold things of certain volume in it, that space is neither shrunk nor expanded nor curved nor moved by any physical means, that the proper speed of light is constant in such absolute space, yet the relative speed of light with respect to observers varies depending on the motion/position of the observers (more on absoluteness of space and the speed of light later).

The reason Einstein recognized ether and the absoluteness of space and time was that he realized that he could not explain the occurrence of centrifugal force (=inertial force), which he identified with gravity in his disc thought experiment. Einstein realized that if space were relative there would be no difference whether or not a disc rotates and that it is illogical that centrifugal force occurs only while a disc rotates and not when the disc does not (see **Chapter 40**). The fact that centrifugal force occurs only while a disc rotates can be explained only by the absoluteness of space.

The uniformness of inertial motion and accelerated motion

(especially of uniform acceleration) call for the absoluteness of space and time because if space and time are not absolute, the concepts of uniform motion and uniform acceleration are impossible or meaningless. Einstein's Leiden speech means that he accepted the classical concept of space and time and straight line.

Strangely enough, however, Einstein soon forgot what he had *repented* at the Leiden University and he returned to his old sinful idea that space and time are relative, shrinkable, expandable, and movable. [5] The idea that space-time is shrinkable, expandable, movable, etc. sounds like the idea that space-time is tangible or material entity. How can space and time, both of which are amorphous and immaterial, form such a tangible entity? Still strange is that modern relativists ignore or forget (intentionally) what their master had repented in his sobriety.

Einstein states in his book Relativity (15th ed.) that the space-time is **Euclidian** in special relativity and that the space-time is **non-Euclidian** in general relativity (see **Chapter 41).** [6] This that space-time is Euclidian when an object is in uniform linear motion and it (space-time) suddenly becomes non-Euclidian if the same object starts being accelerated. What if object *A* is in uniform motion and object *B* is in accelerated motion in a given inertial environment? For Einstein, space-time can be anything in order to support the ***theory*** of isotropy of light speed. Einstein ordered nature to obey his theory. Then Einstein is a law-giver or god. (In theology, only a god gives laws to nature.)

7. Einstein violated the isotropy of light speed

Einstein believed that the speed of light is the same (*c*) for all observers/objects. But he violated the isotropy of light speed in his *train thought experiment*, from which he derived the ***relativity of simultaneity*** (see **Chapter 15**). [7]

The *long train* (the length of the train is not important) in Einstein's thought experiment is in uniform motion. Einstein used to treat the space (room) inside the train as an inertial reference body. Then, the light rays from the front part of the train and the light rays from the rear part of the train should meet the observer at the mid-point of the train at the same time no matter what speed the train moves in whatever direction. Then, Einstein's relativity of simultaneity is wrong.

But Einstein concluded differently; he concluded that the light from the front part of the train meets the observer at the mid-point of the train

Summary

earlier than the light from the rear part of the train does. This means that Einstein violated the isotropy of light speed. In his train thought experiment Einstein killed two birds— the isotropy of light speed and the relativity of simultaneity—at the same time. Congratulations!

Sagnac (1869-1928) disproved the isotropy of light speed by using a rotating interferometer in 1913 (see more in **Chapter 15**). Einstein's train thought experiment mentioned above is a kind of *Sagnac experiment*, so to speak: The train in Einstein's experiment is equivalent to an interferometer in motion on the round surface of the earth (it does not matter whether the surface of the earth is round or flat), and the observer at the midpoint of the train is equivalent to the viewing screen of Sagnac' interferometer. We should note that what Sagnac dealt with in his experiment was the *relative speed* of light with respect to the viewing screen (which is equivalent to an observer) and not the proper speed of light with respect to absolute space. Sagnac, like his contemporaries, did not discern the proper speed of light (with respect to absolute space) from the relative speed of light (with respect to a specific observer).

Einstein violated the isotropy of light speed again in his *chest thought experiment*, from which he derived general relativity (more story in **Chapter 39**). [8] In the chest thought experiment, Einstein states that if a beam of light enters a chest, which is in accelerated motion with respect to *Galilean space* (free space), the path of light looks curved inside the chest (see **Fig. S-3**).

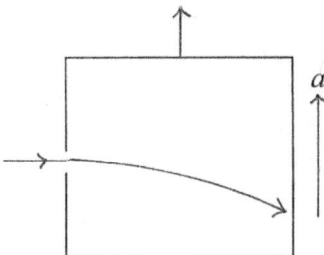

Fig. S-3 A beam of light enters a chest, which is being accelerated. The path of light looks curved when viewed inside the chest.

The fact that the path of light beam curves inside the chest means that the speed (*relative speed*) of light with respect to the chest is changed (continuously)! The chest in Einstein's experiment needs not be in accelerated motion to demonstrate the change of the speed of light.

The path of light will bend (change) if the chest is in *uniform motion* with respect to the Galilean space. In this case the speed of light will change *once*, not continuously as in the case of accelerated motion of the chest, and proceed in linear fashion (see **Fig. 37-4** in **Chapter 37**).

The curving or bending of light beam inside the chest means that the *relative speed* of light with respect to the chest (and the observer inside the chest) changes due to the motion of the chest. But when viewed from the Galilean space (free space) outside the chest, the path of light is straight. That is, the proper speed of light with respect to the free space (absolute space, in author's interpretation) is constant. Einstein acknowledged the change of *light speed* in his chest thought experiment as follows: [9] (The words in the brackets are authors.)

> .... A curvature of rays of light can only take place when the velocity of propagation of light varies with position. Now we might think that as a consequence of this, the special theory of relativity and with it the whole theory of relativity would be laid in the dust. But in reality this is not the case. We can only conclude that the special theory of relativity cannot claim an unlimited domain of validity; its results hold only so long as we are able to disregard the influences of gravitational fields on the phenomena (*e.g.* of light).
>
> --------------------------------------

The curving or bending of light beam inside the chest is simply the phenomenon of ***Newtonian relativity***. And the curving or bending of the light beam inside the chest means that the speed (relative speed) of light with respect to the chest changes. In this thought experiment, what changes is the relative speed of light with respect to the chest in motion and not the proper speed of light with respect to the Galilean space.

8. Relativity of acceleration is wrong.

In the chest thought experiment Einstein though that there occurs gravitational field inside the chest. Einstein identified the inertial force coming from the accelerated motion of the chest with gravitational force. Einstein regarded the chest in accelerated motion as an *inertial frame of reference*. Concerning this Einstein states as follows: [10] (The words in the brackets are authors.)

Summary

> Even though it [the chest] is being accelerated with respect to the 'Galilean space' first considered we can nevertheless regard the chest as being at rest. We have thus good grounds for extending the principle of relativity to include bodies of reference which are accelerated with respect to each other.

The above argument can be referred to as ***the relativity of acceleration***. The author does not recognize this sort of relativity. The reason is as follows:

Suppose that chest A is in stationary state in the Galilean space and chest B is being accelerated with respect to chest A. Or suppose that railcar A is stationary on a straight railroad and railcar B on the same railroad is being accelerated with respect to railcar A. In these cases, we can say that the two chests or the two railcars are in accelerated motion with each other so that Einstein's relativity of acceleration is true.

However, the gravity-like force (inertial force) occurs only in the party (chest B or railcar B) that is *actually* being accelerated; the gravity-like force (=inertial force) does not occur in chest A or railcar A, which is stationary in the given system. In other words, gravity-like force (inertial force) occurs only to the one that is in ***actual accelerated motion*** and not to the one that is in ***virtual accelerated motion***. The state of being *actually accelerated* of an object is the state in which actual force acts continuously on the object. There is no actual force involved in the party that is stationary or *virtually accelerated*. Therefore, the relativity of acceleration is wrong.

9. Is light speed constant for all observers?

We have seen in the previous section that Einstein violated the isotropy of light speed. In this section we will discuss the meaning of the isotropy light speed some more and how Einstein came to believe in the isotropy of light speed.

The isotropy of light speed means that the speed of a (any) photon is the same relative to all human observers and all non-human objects in the world regardless of whether the observers or objects watch (observe) the photon or not.

Observers (and objects) are arbitrary or unpredictable in their actual motion and position. (Note that actual motion/position is different from relative motion/position.) If the relative speed of a photon with respect to any (all) observer/object is to be the same (c), the following should be true:

The number of humans and non-human objects in the world (universe) is gazillion. Each and every photon maintains the same relative speed (c) with respects to all the humans and non-human objects in the universe at the same time. This means that each photon moves at gazillion of different actual velocities (speeds + directions) at the same time. That is, the actual speed of a photon is not constant. Why? Because if a variable $\pm$ actual light speed $= c$ (=constant relative speed of light), the actual speed of the light should be variable.

On the other hand, if we are to think of the reciprocity of relative speed, the isotropy of light speed means that all observers or objects in the whole universe move at the speed of light (c) relative to each and every photon.

The above stories, which are simply nonsense, are the foundation of the theory of relativity. But Einstein succeeded in making such an impossible theory a truth. How? He borrowed the idea of Lorentz that space and time orchestrate and weave four-dimensional magical fabric that expands, contracts, curves, wrinkles, or whatever in order to make the speed of light always the same relative to all observers/objects. The theory of relativity is a delusion.

Even before Einstein published his special relativity (1905), many of his contemporaries and predecessors had believed in the isotropy of light speed and absoluteness of space allegedly filled with material ether. But they had not distinguished the concept of relative speed from that of proper speed. Einstein did not either.

According to the Einstein's biography written by Walter Isaacson, Einstein had agonized a lot at first because he could not understand why or how the speed of light is always the same for all observers regardless of the different state of motion of the observers. [11] In this respect, Einstein was inquisitive in the right direction at first. But he quit his inquisition; he just accepted the then cutting-edge theory of Lorentz, who first explained the isotropy of light speed with a set of equations called the *Lorentz transformation* (LT) (more on this in **Chapter 24**). It should be noted that Lorentz believed in ether and absolute space and he did not distinguish proper speed from relative speed.

Summary

The Lorentz transformation (LT) and the relativistic formula of speed addition/subtraction, the latter of which is the derivation of the former, surely prove the isotropy of light speed, at least mathematically.

$$\frac{c \pm v_{any}}{1 + \frac{c \times (\pm v_{any})}{c^2}} = \frac{c \pm v_{any}}{\frac{c(c \pm v_{any})}{c^2}} = c.$$

(c is the speed of light, and v_{any} is the speed of any observer/object.)

We need not marvel at the magical power of the speed addition/subtraction formula because this formula was forged for the purpose of making the speed of light always the same (c). For example, the above expression holds even if we replace v_{any} with one thousand times the speed of light ($1000c$) or one trillion times the speed of light ($1,000,000,000,000c$). For example,

$$\frac{c \pm 1000c}{1 + \frac{c \times (\pm 1000c)}{c^2}} = c. \text{ (see the detail in \textbf{Chapter 23})}$$

Therefore, the act of proving or supporting the isotropy of light speed using the relativistic formula of speed addition/subtraction is but ***circular reasoning***. In addition, we should note that the Lorentz transformation and the addition of speed is a 1-D (one-dimensional) manipulation because Lorentz considered speed based on the assumption that two involved objects (a photon and an observer or origin, for example) are situated or moving on the same straight line (more on this in **Chapter 24**).

10. Relative to what is light speed constant?

The propagation of light in space is a physical phenomenon. If light speed is constant, it should be so with respect to a certain *environmental reference body* and not to a *point reference body* such as a human observer or other object (see the concepts of *environmental reference body* and *point reference body* in the third section of Summary). In other words, the ***proper speed*** of light with respect to a certain environmental reference body (space) should be constant; the *relative speed* of light with respect to point reference bodies (observers) cannot be constant.

Having considered many fallacies or problems with relativity the author came to the conclusion that space is absolute without ether (see

Chapters 18 and **21**) and that the proper speed of light is constant with respect to absolute space and not with respect to human observers or other objects whose motion and position are arbitrary and unpredictable.

The author thinks that space and time are absolute and independent of each other. This is Newton's concept of space and time. The first law of motion (the law of inertia) and the second law of motion (the relationship between force, mass, and acceleration) of Newton have to do with the absoluteness of space and time. For example, inertial motion is uniform with respect to absolute space and absolute time.

Inertial motion—a kind of uniform linear motion—in space is meaningful only when both space and time are absolute. The concept of *uniform motion* or uniform acceleration is meaningful only when space and time is absolute. I wonder how Einstein had the thought of using such terms or concepts like uniform motion, inertial coordinate system, etc. without recognizing the absoluteness of space and time.

The reason Einstein belatedly recognized ether and the absoluteness of space and time in his Leiden address (see **section #6**) was that he realized that the occurrence of inertial force (from accelerated motion) in his chest thought experiment and disc thought experiment is possible only when space and time are absolute. [12] [13] But Einstein soon forgot what he had said and returned to his former idea that space and time are relative.

Einstein's concept of space was inconsistent; he continued modifying his concept of space into the later years of his life. Einstein added a *note* titled "**Note to the Fifteenth Edition**" after the preface of his book *Relativity* (15th ed., 1952) as follows. [14]

Note to the Fifteenth Edition

In this edition I have added, as a fifth appendix, a presentation of my views on the problem of space in general and on the gradual modifications of our ideas on space resulting from the influence of the relativistic view-point. I wished to show that space-time is not necessarily something to which one can ascribe a separate existence, independently of the actual objects of physical reality. Physical objects are not *in space*, but these objects are *spatially extended*. In this way the concept "empty space" loses its meaning.

A. Einstein   June 9th, 1952

Summary

In the fifth appendix titled "Relativity and the Problem of Space" in the same book Einstein comments on the concept of Newtonian space and that of Cartesian space as follows: [15]

> It is characteristic of Newtonian physics that it has to ascribe independent and real existence to space and time as well as to matter, for in Newton's law of motion the idea of acceleration appears. But in this story, acceleration can only denote "acceleration with respect to space." Newton's space must thus be thought of as "at rest," or at least as "unaccelerated," in order that one can consider the acceleration, which appears in the law of motion, as being a magnitude with any meaning. Much the same holds with time, which of course likewise enters into the concept of acceleration. Newton himself and his most critical contemporaries felt it to be disturbing that one had to ascribe physical realities both to space itself as well as to its state of motion; but there was at that time no other alternative, if one wished to ascribe to mechanics a clear meaning.
>
> It is indeed an exacting requirement to have to ascribe physical reality to space in general, and especially to empty space. Time and again since remotest times philosophers have resisted such a presumption. Descartes argued somewhat on these lines: space is identical with extension, but extension is connected with bodies; thus there is no space without bodies and hence no empty space. The weakness of this argument lies primarily in what follows. It is certainly true that the concept of extension owes its origin to our experiences of laying out or bringing into contact solid bodies. But from this it cannot be concluded that the concept of extension may not be justified in cases which have not themselves given rise to the formation of this concept. Such an enlargement of concepts can be justified indirectly by its value for the comprehension of empirical results. The assertion that extension is confined to bodies is therefore of itself certainly unfounded. We shall see later, however, that the general theory of relativity confirms Descartes' conception in a roundabout way. What brought Descartes to his remarkably attractive view was certainly the feeling that, without compelling necessity, one ought not to ascribe reality to a thing like space, which is not capable of being "directly experienced."

Einstein's later concept of space was basically that of Descartes.

The author finds no problem with Newton's concepts of space, time, inertia, acceleration, and laws of motion. The author explains (affirms) the absoluteness of space as follows:

Suppose that a spaceship is in inertial state in free space. There are no sizable bodies in the vicinity of the spaceship to be seen or to influence the state of motion of the spaceship in terms of gravitational field. In this case, how do we know whether the spaceship is in *inertial state* in free space? We or the crew inside the spaceship can confirm the inertial state of the spaceship by checking that there are no gravity-like effects or inertial force in any direction inside the spaceship. Yet we still do not know whether the spaceship is in stationary state or in uniform inertial motion with respect to space.

Let us just suppose that the spaceship is stationary with respect to space. Now, let us suppose again that the pilot of the spaceship turns on the booster rocket engine, which is designed to accelerate the spaceship linearly in its head direction; the engine is on for five seconds (the length of time is not important) and then it is turned off. Now, we can say that the spaceship has been accelerated for five seconds with respect to space and that the spaceship is now in uniform linear motion at a certain speed in a certain direction in space. We might be able to calculate the proper speed of the spacecraft in absolute space or the distance the spacecraft covers per second by taking account of the power of the boosting engine, the time (duration) the engine has been on, the total mass of the spaceship, the rate of the decrease of the mass of the spacecraft due to the discharging of the burnt fuel, etc. (But calculating the speed of the spacecraft is not our concern in this thought experiment.) All these stories sound logical and natural if space and time are absolute.

Now, the author asks a question to relativists or readers concerning the above thought experiment: Is the position of the spaceship at the moment the engine was started the same as the position of the spaceship at the moment the engine was turned off in five seconds? If space is absolute, our answer would be clear and easy: "The two positions of the spaceship are not the same with each other." But for relativists, who negate the absoluteness of space and time, the answer is not available.

Einstein was uneasy with the concept of empty space; he denied empty space. Einstein defined the position of an object or event **with respect to a rigid body** (another object). Einstein states in his book

Summary

Relativity (15th ed.) as follows: [16]

> Every description of the scene of an event or of the position of an object in space is based on the specification of the point of a rigid body (body of reference) with which that event or object coincides. This applies not only to scientific description, but also to everyday life.

In the author's spaceship thought experiment above, can we know whether the spaceship is stationary in the absolute space or whether it moves at a certain speed and direction with respect to space? The author thinks we can. For example, we put a light source, which is designed to emit regular intermittent light pulses, on the nose (or any other part) of a spaceship. If the spaceship is stationary relative to absolute space, the light pulses would make *concentric circles* around the source as shown in diagram (a) in **Fig. S-4**. However, if the spacecraft is in motion, the light pulses would make *eccentric circles* as shown in diagram (b).

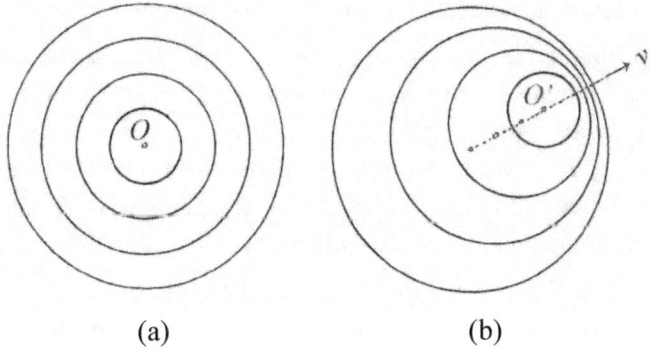

(a)   (b)

Fig. S-4 Source O in (a) is stationary. Source O' in (b) is in motion.

By analyzing the eccentric circles, in the case of (b), we might be able to find the absolute velocity (speed + direction) of the spaceship in absolute space. This is a theoretical method. The problem is that we have no means to gather and analyze the information of the eccentric or concentric circles of light pulse on *then-and-there* basis.

For relativists, who believe in the isotropy of light speed, eccentric circles of light pulse do not occur. Arthur Beiser explains in his text book

Concepts of Modern Physics that light waves propagate always at the same speed (*c*) from the source or observer regardless of whether the source (or observer) moves or not.[17] Doppler Effects of light cannot not occur! Then, it is contradictory that Beiser, explains (recognizes) the Doppler Effect of light in the same book. [18]

11. Is Absolute Coordinate System Possible?

In the previous section we have discussed the absoluteness of space and time. Some readers, if they acknowledge the absoluteness of space and time, might think we may be able to set the absolute coordinate system so that we can give absolute coordinate values to any heavenly body at a specific time point by making the most of fixed stars. But the author thinks we cannot. Most fixed stars are not perfectly fixed. In addition, the author proved *mathematically* that the relative speed of the earth (or any object) with respect to a (any) fixed star is **zero** (0) or almost zero in spite of the absolute motion of the earth in space (see **Chapter 6**). Therefore, we cannot give the absolute coordinate values to a point (object) in space at a certain time with respect to fixed stars. But the fact that we cannot give specific coordinate values to a point in space does not disrupt the absoluteness of space and time.

According to Newton's concept of space, the position of the earth in space at noon time yesterday is different from the position of the earth at noon time today even though we cannot give specific coordinate values (such as x, y, z, t) to the position of the earth in space. But relativists cannot tell whether the position of the earth yesterday is the same as or different from the position of the earth today.

In GPS (global positioning system), relativists assume the earth as a stationary reference body (this is somewhat like the idea of geocentricism). According to relativity, if an observer on the earth sends out light or electromagnetic signals into space, the speed and direction of such signals is always the same with respect to the earth (or the observer on the earth). If our earth is assumed to be always stationary with respect to space, **relativistic information correction** of GPS might be true (see **Chapter 14**). But in fact our earth is neither in stationary state nor in inertial state. The earth is in complicated motion consists of rotation motion and circular motion (revolution) in space.

Relativists argue that GPS is correct due to the relativistic correction. Relativists' argument is not true: GPS does not depend on the theory of

Summary

relativity; relativists wrongly apply relativity to GPS (see **Chapter 14**). The application of relativity to GPS is to put a fifth wheel to a cart.

Readers might ask why modern physics is not crumbling before our eyes if relativity is wrong. That is because the *alleged* relativistic effects, most of which fall in the range of experimental errors, are so small and negligible that they do not affect our everyday lives.

12. One-way light speed is not constant.

Here is sad news for relativists concerning the isotropy of light speed: According to the latest method of measuring ***one-way light speed***, the speed of light (=the *relative speed* of light measured in one-way method) turned out to be not constant ($\neq c$). [19] The conventional method of measuring light speed was two-way average method. If relativity were true, the light speed should always be the same (c) whether it is measured in two-way average method or one-way method (more on this in **Chapter 21**). The outcome that one-way light speed was not constant was disappointing for the experimenters who had expected the isotropy of light speed in any case.

The above result (the ***inconsistency*** of one-way light speed) can be understood if we distinguish the ***proper speed*** of light with respect to absolute space) from the ***relative speed*** of light with respect to light-speed measuring apparatus. The author thinks that the cause of "the inconsistency of one-way light speed" was due to the complicated motion of the earth (and the light-speed measuring apparatus on the earth) with respect to absolute space. The absolute speed of the light measuring apparatus on the earth varies irregularly due to the irregular motion of the earth with respect to absolute space. The author also thinks that the presence of the atmosphere or air flow (wind) might have contributed minor, if not major, influence.

Relativists, who believe in the Lorentz math (such as $\sqrt{1 - v^2/c^2}$), persist that light speed (c =300,000km/s) is the ***upper limit of speed***, without thinking whether c is the proper speed or relative speed. When it comes to relative speed, the speed of a photon with respect to a certain object (including another photon) can be from zero (0) to $2c$. For example, if two photons, say A and B, move in the same direction on the same straight line, the relative speed between A and B is $c \sim c = 0$. If A and B move in the opposite direction of each other on the same straight line, the relative speed between the two is $c \sim (-c) = 2c$. (Note that the sign of the light speed of B in this case is minus because the direction of motion of B is opposite of that of A.)

Relativists might not recognize the varying values of relative speed of light (0 or $2c$) for the reason that these values are *virtual speeds* and not actual speeds. Yes, these speeds are virtual speed of light. But relativists should note that the theory of relativity is about relative speed, which is either actual or virtual, and not about actual speed (see the concepts of actual speed and virtual speed in **section 4**). Therefore, it is wrong for relativists to disregard virtual speed of light.

When it comes to relative speed, a human observer can move at the speed of light (c) or faster-than-light speed ($c + v$) with respect to a photon. This kind of speed is virtual one and not an actual one. The theory of relativity and concept of the isotropy of light speed do not distinguishing actual speed from virtual speed.

13. The law of the equality of gravitational mass and inertial mass is wrong.

Einstein declared *the law of the equality of inertial and gravitational mass* in his chest thought experiment as follows: [20] (The words in the brackets are the authors.)

> We must note carefully that the possibility of this mode of interpretation [= the state of being accelerated is the same as the state of being at rest] rests on the fundamental property of the gravitational field of giving all bodies the same acceleration, or, what comes to the same thing, on the law of the equality of inertial and gravitational mass. If this natural law did not exist, the man in the accelerated chest would not be able to interpret the behavior of the bodies around him on the supposition of gravitational field, and he would not be justified on the grounds of experience in supporting his reference-body to be "at rest."
> -------------------------------------

Einstein keeps blundering. Inertial mass is not the same as gravitational mass. The inertial mass of an object is defined by the Newton's second law of motion, which is expressed by a simple equation $F = ma$, where F is force, m is mass (inertial mass), and a is acceleration. This means that, for example, *one Newton* of force acting on 1 kilogram of mass (inertial mass) can change its velocity by 1 meter per second in the direction of the force. [21]

We usually refer to gravitational mass as *weight*. Weight or gravitational mass of an object is determined by the magnitude of the gravitational force of a planet on which the object rests. The weight of an object becomes smaller if the object is moved away from the surface of the planet (because the gravitational force diminishes if the distance of the object from the center of gravity of the planet increases). [21] The weight of an object also depends on the mass of the planet. For example, a man who weighs 200 pounds (91kg) on the earth would weigh only 32 pounds (15kg) on the moon [22]. These are what high school students learn and understand. Since Einstein identified inertial force with gravitational force (see next section), he could not but come to the wrong conclusion that inertial mass is the same as gravitational mass.

14. Inertial force is not the same as gravity (the equivalence principle is wrong).

The core of the general theory of relativity is that the inertial force, which comes from acceleration, is the same as gravity, which comes from the body of matter. Einstein called this ***the principle of equivalence*** or ***equivalence principle***. [23]

Einstein asserts in his ***chest thought experiment*** that the inertial force inside the accelerated chest is the same as gravity. [23]

Einstein asserted in his *disc though experiment* that the centrifugal force (=inertial force) coming from the rotation motion of the disc is the same as gravity, and he gave an annotation in his book as follows: (The words in the brackets are authors.) [24]

> "The field [gravitational field] disappears at the center of the disc and increases proportionately to the distance from the center as we proceed outward."

In the above annotation Einstein is talking about centrifugal force, which is inertial force.

Relativists assert that *the effect of inertial force* is the same as gravity so that they cannot distinguish the former from the latter. But the author has found at least *eight different characteristics* that clearly and readily distinguish inertial force from gravity. The eight different characteristics are, without the order of importance, as follows:

#1. Gravitational force from the body of matter lasts (or seems to last) forever as long as the mass of the body is preserved. But inertial force coming from acceleration lasts only while an object is being accelerated.

#2. The distance gravitational force reaches is limitless. But the inertial force is limited to the body (laboratory) being accelerated.

#3. The magnitude of gravitational force diminishes as the distance from the center of gravity grows. But the magnitude of inertial force of an accelerated laboratory is the same everywhere inside the laboratory. In the case of rotation motion, the centrifugal force (=inertial force) grows as the distance from the center of the rotation grows.

#4. There is no known means to block or adjust the gravitational force from the matter of body. But the direction and magnitude of inertial force of an accelerated body can be readily controlled or adjusted. For example, the direction and size of inertial force inside a spacecraft are readily controlled by adjusting the power and direction of the engines.

#5. The lines of force of gravity spread radially from the center of gravity of an object as the spines of a sea urchin spread out from the body of the animal. But the lines of inertial force from acceleration are parallel in the direction of the acceleration.

#6. The acceleration of free-falling caused by natural gravity is uniformly maintained. For example, the acceleration of a free-falling body in the earth gravitational field is uniform ($9.8 m/s^2$) (if we ignore the resistance of air). But the artificial acceleration of an elevator or chest, for example, is not always or necessarily uniform. Einstein thought only of the case of uniform acceleration and uniform rotation. [25]

#7. Gravity is explained by Newton's law of gravity whereas inertial force is explained by Newton's first law of motion (law of inertia). The law of gravity is different from the law of inertia.

#8. Gravity is one of four *real forces* that occur in nature. The four real forces are (1) gravity, (2) electromagnetic force, (3) weak nuclear force, and (4) strong nuclear force. In contrast, inertial force is pseudo (false) force; inertial force is the resistance against the change of speed or direction of motion.

Summary

We do not need any more different characteristics to distinguish gravity from inertial force. The nature of gravity from the body of matter is still mystery. We, at least the author, do not fully understand the nature of gravity. But inertial force is fully and clearly explained by Newton's law of motion.

Einstein thought only of uniform inertial force from *uniform acceleration*. But inertial force is not always or necessarily uniform. Inertial force can be uniform, non-uniform, or one-time abrupt jerk.

Some readers might argue that the free-falling motion of an object in the earth's gravitational field is a case of uniform acceleration and that this is the same as the uniform artificial acceleration of a laboratory (chest, spaceship, etc.). They are wrong; there is a fundamental difference between the two cases: The gravity-like inertial force occurs only in the chest that is being accelerated *artificially*: There is no gravity-like inertial force inside the free-falling chest in the natural gravitational field. Gravity is the cause of acceleration, not inertial force. Artificial acceleration is the cause of inertial force, not gravity. Artificial gravity (inertial force) is fundamentally different from natural gravity.

15. Light does not curve in the gravitational field.

Einstein concluded in his chest thought experiment that the inertial force, which comes from the acceleration of the chest, is the same as gravity. Einstein further reasoned that the curving of light beam inside the chest is due to the gravitational field inside the chest. Einstein was wrong: The curving of the light beam inside the chest is simply the phenomenon of *Newtonian relativity* (see **section # 7**).

Based on the idea that light curves in the gravitational field inside the accelerated chest, Einstein predicted that starlight should deflect at a certain degree (1.74 arc second; one arc second is a 3600th second) when it passes through the gravitational field of the sun. Einstein's reasoning is wrong because inertial force is not the same as gravity.

Relativists think that gravitational lens effect is due to the curving of light in the gravitational field of a large planet. But the author thinks that gravitational lens effect is due to the large body of cosmic gases around some (not all!) heavenly bodies. If gravitational lens effect is due to gravity of stars, such effects should be found around any sizable heavenly bodies (more on this in **Chapter 37**).

16. Eddington's eclipse observation was a sham.

Relativists believe that Eddington proved the principle of equivalence or general relativity through his eclipse observation in 1919. But we have already seen in the previous sections that the equivalence principle is wrong and that the curving of light beam inside an accelerated chest is simply the phenomenon of Newtonian relativity and not because of gravity.

Aside from the verity or falsity of the equivalence principle or general relativity, the author proves that Eddington's "method of eclipse observation" itself was a sham with the following reasons:

#1. Eddington had already been an avid believer of the theory of relativity even before he planned the eclipse observation. Eddington's costly and grand-scale observation expedition was to *confirm* Einstein's theory rather than check it objectively and impartially. Eddington was the director and supervisor of the entire project of the eclipse observation. He was the sole *commander-in-chief*, so to speak, in the eclipse observation expedition. These elements show that Eddington was not the proper person to perform such an important observation in the first place.

#2. Eddington and Einstein were in a special relationship; both were pacifists and socialists. Due to their common ideology, which was popular among many intellectuals in Europe during the time of WWI, both Eddington and Einstein were criticized in their own countries respectively. Such a circumstance must have helped the two people seek some sort of cooperation or solidarity.

#3. The lineup of the star, sun, moon, and earth during the time of the eclipse was fragile one and thus did not last long. The *quality* of such celestial lineup of the heavenly bodies changes in a matter of a second or fraction of a second. We have the experience of watching the setting sun sink below the horizon at a considerable (noticeable) speed even with our unaided eye. Eddington's photos must have recorded a series of the moments *of fast-changing line-up* of the sun, moon, earth, and the star. These photos were taken by old-fashioned manual cameras of the early 20th century technology. There was no powerful telescope or computer. It is not hard to imagine that Eddington could have picked and chose some "proper" photos among the serial photos in order to support the theory he already firmly believed in.

#4. Eddington had known the *predicted value* of the starlight

Summary

deflection (1.7 arc-seconds) even before he planned the eclipse observation. This is somewhat like the situation that a student knows the correct answer of a problem before he takes an exam.

#5. Einstein's prediction of starlight deflection around the sun was not his original idea; Einstein must have had fiddled with Newton's theory of starlight deflection due to sun's gravity. In his 1911 paper, Einstein predicted that the deflection angle would be 0.85 arc-seconds. This value was exactly the same as what Newton predicted based on his *corpuscular theory*. [26] The author does not think Einstein's prediction (0.85 arc-second) was an innocent coincidence. How could the two entirely different theories end up with the same deflection angle? One year later, in 1912, however, Einstein corrected his calculation from 0.85 arc-seconds to 1.7 arc-seconds, exactly two times the calculation of Newton. [26] Still Einstein is not immune from our suspicion of his *mathematical fiddling*. We, at least the author, do not know whether Newton's prediction was/is true.

#6. Eddington and Einstein did not take the thick and turbulent *solar atmosphere* into consideration. Since the solar atmosphere is from several hundreds to thousands miles thick and very turbulent, it could easily override the starlight deflection or diffraction, if any. It is said that scientists in the time of Einstein and Eddington, let alone Newton, did not know of the presence of the thick and turbulent solar atmosphere.

#7. The size of the images of the star captured in Eddington's photos must have been very small but not small enough to be considered a *mathematical point*. And the boundary of the image of the sun on the photos must have not been so sharp and clear as that of mathematical line. The job of measuring, calculating, and analyzing the value of the deflection of the starlight was done manually by Eddington alone. We might well suspect that Eddington's manual job with not-so-sharp images must have incurred some degree of innocent or intentional or subconscious *human error*.

#8. Historical records tell us that the output of the measurements and analysis of Eddington were not proper to be considered the proof of general relativity. But Eddington pushed his opinion in the discussion, in which Eddington himself was the sole expert in relativity; and his opinion was chosen as the official proof of Einstein's theory (more stories in **Chapter 42**). [27]

Even if Eddington's analysis (data) of the eclipse photos was correct and honest, the data cannot prove general relativity because we have already seen in the previous sections that the equivalence principle is wrong and that light does not deflect in the gravitational field. If one gets correct data from his observation/experiment, which is based on wrong principle, it means that his observation/experiment is a sham.

17. Gravity vs. geometry (curvature) of space-time

Einstein negated the classical concept of gravity and replaced it with the geometry (curvature) of space-time. Arthur Beiser illustrates the geometrical curvature of space-time in his textbook as follows: [28]

> General relativity pictures gravity as a warping of space-time due to the presence of a body of matter. An object nearby experiences an attractive force as a result of this distortion, much as a marble rolls toward the bottom of a depression in a rubber sheet. To paraphrase J. A. Wheeler, space-time tells mass how to move, and mass tells space-time how to curve.

Even if we concede that the slope or depression of space-time is formed around the body of matter, the slope of the space-time fabric alone can make an object nearby roll down the slope however steep the slope may be unless there is attracting force from the body of matter.

People used to think that objects roll or slide down the slope from higher place to lower place. This is due to the presence of earth's gravity. If there is no gravity as an attracting force, objects do not roll or slide down the slope. Space-time depression model lacks the major factor—the attraction force. Newtonian concept of gravity is natural, simple, and direct and thus superior to Einsteinian space-time slope model.

18. $E = mc^2$

The author negates the Einstein's energy equation ($E = mc^2$) simply as follows:

The speed of light (c) in the energy equation is neither proper speed nor relative speed. There is no such light speed that is the same for all observers and non-human counterpart objects. It is said that the energy equation was formulated not by Einstein but other scientists. Those "other scientists" did not discern relative speed from proper speed of light either.

Summary

Most people think that the energy equation has contributed to the invention of atomic bomb. They are wrong; Einstein's proposal (the letter to then President Roosevelt) could have helped in political manner in decision making of the manufacture of the atomic bomb. But the energy equation itself was/is far from the principle of atomic bomb (more on this in **Chapter 32**). [29] According to relativity, mass (m) and energy are relative quantity depending on the motion/speed of observers (or counterpart objects). The author does not recognize the energy equation.

19. Einstein, the great plagiarist

Einstein's method of studying science was not honest; he borrowed most ideas from his contemporaries or predecessors without proper attribution. The website article entitled "Einstein, Plagiarist of the Century" states as follows: [30]

> Proponents of Einstein have acted in a way that appears to corrupt the historical record.
>
> Albert Einstein (1879 -1955), Time Magazine's "Person of the Century", wrote a long treatise on special relativity theory (it was actually called "*On the Electrodynamics of Moving Bodies*", 1905a [sic]), without listing any references. Many of the key ideas it presented were known to Lorentz (for example, the *Lorentz transformation*) and Poincaré before Einstein wrote the famous 1905 paper.
>
> As was typical of Einstein, he did not discover theories; he merely commandeered them. He took an existing body of knowledge, picked and chose the ideas he liked, then wove them into a tale about his contribution to special relativity. This was done with the full knowledge and consent of many of his peers, such as the editors at *Annalen der Physik*.
>
> The most recognizable equation of all time is $E = mc^2$. It is attributed by convention to be the sole province of Albert Einstein (1905). However, the conversion of matter into energy and energy into matter was known to Sir Isaac Newton ("*Gross bodies and light are convertible into one another...*," 1704). The equation can be attributed to S. Tolver Preston (1875), to Jules Henri Poincaré (1900; according to Brown, 1967) and to Olinto De Pretto (1904) before Einstein. Since Einstein never correctly derived $E = mc^2$ (Ives, 1952), there appears nothing to connect the equation with anything original by Einstein.

> Arthur Eddington's selective presentation of data from the 1919 Eclipse so that it supposedly supported "Einstein's" general relativity theory is surely one of the biggest scientific hoaxes of the 20th century. His lavish support of Einstein corrupted the course of history. Eddington was less interested in testing a theory than he was in crowning Einstein the king of science.
>
> The physics community, unwittingly perhaps, has engaged in a kind of fraud and silent conspiracy; this is the byproduct of simply being bystanders as the hyperinflation of Einstein's record and reputation took place. This silence benefited anyone supporting Einstein. ....
>
> -----------------------------------

Christopher Jon Bjerknes writes in his article "A Theory of Einstein the Irrational Plagiarist," as follows: [31] (This article was published in *The Canberra Times, September 19, 2006*.)

> The completed field equations of the general theory of relativity were first deduced by David Hilbert, a fact Einstein was forced to acknowledge in 1916, after he had plagiarised them from Hilbert in late 1915. Paul Gerber solved the problem of the perihelion of Mercury in 1898. Physicist Ernst Gehrcke gave a lecture on the theory of relativity in the Berlin Philharmonic on August 24, 1920, and publicly confronted Einstein, who was in attendance, with Einstein's plagiarism of Lorentz' mathematical formalisms of the special theory of relativity, Palagyi's space-time concepts, Varicak's non-Euclidean geometry and of the plagiarism of the mathematical solution of the problem of the perihelion of Mercury first arrived at by Gerber. Gehrcke addressed Einstein to his face and told the crowd that the emperor had no clothes. ....
>
> Numerous republished quotations from Einstein's contemporaries prove that they were aware of his plagiarism. Side-by-side comparisons of Einstein's words juxtaposed to those of his predecessors prove the almost verbatim repetition. There is even substantial evidence presented in the book that Einstein plagiarised the work of his first wife, Mileva Maric, who had plagiarised others.
>
> -----------------------------------

Relativists defend Einstein by saying that there are none among scientists who do not borrow the ideas or knowledge of their peers or

Summary

predecessors. But borrowing other's ideas without proper attribution is different from doing so with proper attribution. What is more, the ideas Einstein borrowed were mostly wrong or unphysical ones; these were based upon the wrong idea of the isotropy of light speed, which came from the ignorance of what relative speed is. The theory relativity is no longer the honor of Einstein or humanity. The theory of relativity has turned out to be the shame for Einstein, his followers, and all humanity.

20. The evidence that supports relativity

If people are proselytized by a certain doctrine or ideology, they can explain all the phenomena under the sun or above the sun with their doctrine or ideology. It is a kind of cult or religious phenomenon. If a group of people believe in a certain idea such as Marxism, or geocentricism or aliens, they *create* tons of observational or experimental data that support their belief. Humans *create* evidence for what they believe or want to support. Believers invent evidence or truth rather than discover them. History shows us ample examples. Sun revolved around the earth before and even after Galileo proved it wrong. Relativity is kind of ***photocentricism*** or the 20th century-version of geocentricism. Relativity is a cult or religion, and Einstein was the law-giver—a god. Einstein ordered nature to follow his law (relativity).

Demons or fairies or aliens really exist for those who believe in these. Four-dimensional space-time continuum, dark matter, dark energy, black hole, gravity lens, worm-hole, the Big Bang, time travel, etc. are the 20th century-version of fairy tales.

Relativists, let alone lay people, do not know what relative speed really is. Not knowing what relative speed is is not an obstacle to become a believer of relativity. Not knowing what relative speed is rather helps or intensifies the belief in relativity.

There was a rumor in the time of Einstein and Eddington that there were only three people in the world who understood the theory of relativity. [32] Einstein and Eddington were the two of the three. The three people, who the third person might have been, did not understand relativity in fact. If anyone thinks he understands the theory of relativity, he does not understand the theory because the theory itself is wrong. Relativity is the confession of misbelievers that they do not know what relative speed is.

Einstein's concepts of speed, space, time, inertia, inertial frame of reference, gravity, inertial mass, gravitational mass, etc. were all wrong. Using tons of misconceptions as the privilege or almighty power, Einstein and his followers have devastated physical science in the name of *modern physics*.

Einstein was inconsistent and wrong in his thoughts not only in physical science but also in the thoughts about religion and philosophy (see **Appendix-3**). The theory of relativity has damaged not only physical science but also other areas such as philosophy, ethics, morality, justice, aesthetics, etc. Many people who are influenced by the theory of relativity think that the concepts of good and evil or right or wrong are the matter of the perspective of viewers. They are wrong. The author has dealt with these themes in other books (not published yet).

21. Relativity trial

Relativity has hampered the development of physical science for over a century (1905-2014): Had it not been the theory relativity, physical science might have achieved one hundred years of advancement and innovation. What a loss! Relativity is the same of humanity. Relativity is more than a shame. Judging from the results, though it was not intentional, relativity is a colossal *crime*: Relativity has fooled humanity and cost billions of tax dollars and incalculable amount of man power.

Yet relativists are adamantly organized never acknowledge the fallacy of relativity. Relativists are equipped with sophistry, abstruse math, political power, money, media, and huge mass of lay people who blindly support the theory of relativity. Relativity cannot be fixed by normal academic activities.

The only solution to fix relativity is to bring it to justice. It takes only a speed gun or two to disprove relativity experimentally in the trial, and the truth will out in seconds. The historical *relativity trial* should be held and aired live to the whole world so that the whole humanity can celebrate the end of the notorious hoax—relativity. ♦

Part I

Special Relativity

According to special relativity, the speed of a photon (light) is the same (c) for all observers/objects regardless of the arbitrary and unpredictable nature of the velocity (speed and direction) of the observers. This means, conversely speaking, that the velocity (speed + direction) of a (any) photon is ***infinitely variable and thus not constant***. Logically speaking, if the sum of a variable plus unknown is constant, the unknown should be another variable. The isotropy of light speed came from confusing relative speed with proper speed.

1

Distance, Speed, and Velocity

Distance

Distance is best defined between two points. In mathematics, a point has only a position and neither size (width, height, depth) nor weight/mass. In physics or reality, distance is defined between two objects each of which can be treated as a point. For example, when we talk about the distance between the earth and the sun, we treat these heavenly bodies as points. If not, it is impossible to determine the distance between two fairly voluminous objects which are apparently not points.

In mathematics, distance is also defined between a point and a straight line, between a point and a plane, between two parallel lines, or between two parallel planes. Even in each of these cases, we measure the distance between two "specific" points on the two involved parties.

We do not define the distance between an object (point, line, or plane) and a body of *space* because space is not represented by a point.

Velocity and Speed

Speed is the rate of change of position of an object. Speed is *scalar* that has *no direction* whereas velocity is *vector* that has a direction. That is, velocity is *speed* plus (+) *direction*. For example, 60km/h is speed; 60km/h, south, is velocity. However, speed and velocity are used interchangeably in everyday lives as long as the direction of motion is not an important issue.

Relative motion/speed is *reciprocal*. For example, the speed of the car in **Fig. 1-1** (this is the same as **Fig. C** on the cover of this book) relative to Einstein is v, and vice versa. But relative velocity (speed + direction) is not reciprocal because the direction of relative motion of one party is different from (opposite of) the direction of motion of the counterpart object. For example, the *direction* of relative motion of Einstein in **Fig. 1.1** is the opposite of the direction of the car with respect to Einstein.

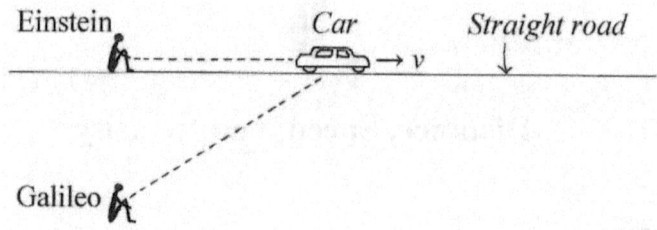

Fig. 1-1 Einstein sits on the road, and Galileo sits off the road.

In the case of the car and Galileo in above situation, the direction of the car with respect to Galileo *changes continuously* with the passage of time. But when it comes to *instantaneous relative velocity* between the car and Galileo, the ***direction*** of *the instantaneous relative motion* of the car with respect to Galileo is the *opposite of* the direction of the instantaneous relative motion of Galileo with respect to the car. Therefore, we can see that *relative speed* is reciprocal but *relative velocity* is not reciprocal.

What is important in relativity is the *absolute magnitude* (positive number) of the rate of change of distance between two objects and not the *direction of motion*. Therefore, it is reasonable and convenient to use the term *speed* rather than *velocity*. For this reason I use mainly *speed* (instead of velocity) in this book unless the *direction* of motion is an important issue.

Proper Speed and Relative Speed

Proper speed (my term) is the distance a *uniformly-moving* object or physical phenomenon (such as light or sound) covers per given unit time in a given inertial environment/system (my definition). By nature, proper speed is independent of the motion/position of any observers. That is, proper speed is *observer-free* or *observer-independent*. For instance, the proper speed of sound in dry-and-*calm* air of 20° C is 343.2 metres per second irrespectively of the motion of the source or observers [1] (The term *calm* means *inertial*.) In **Fig. 1-1**, the proper speed of the car is v

Special Relativity

and constant independently of the motion/position of Einstein, Galileo, or any other observers/objects.

The method of measuring the proper speed of a uniformly-moving object in a given inertial environment is as follows: We measure the distance the object has covered in a certain period of time, and we divide the distance by the time taken. Another method is that an observer sits on the line of motion of the object, and the observer aims a speed gun at the target object which is in uniform linear motion along its line of motion. Proper speed is one dimensional in concept because it occurs on a straight line.

Relative speed is the *rate of change in the distance between two specific objects* (my definition). Even if an object is in uniform linear motion in a given inertial environment/system, the relative speed of the object with respect to a specific observer may vary depending on the motion/position of observer. For example, the relative speed of the car in **Fig. 1-1** with respect to Einstein is v. This value is the same as the proper speed of the car because Einstein sits on the line of motion of the car. But the relative speed of the car with respect to Galileo is less than v and changes continuously as the car moves. This phenomenon, which has nothing to do with relativity, is consistently confirmed by calculus, trigonometry (cosine effect), and radar (speed detector or speed gun) (see **Chapters 4-8**).

Practically all physicists (both relativists and dissenters) I have contacted so far (2001~) do not distinguish the concept of relative speed and that of proper speed. They do not understand that the relative speed between Galileo and the car is less than v and changes continuously. This is because physicists have habitually and customarily considered speed only in 1-D situations or in terms of 1-D (one-dimensional) situation even when the actual situation is 2-D or 3-D.

The 1-D situation is the case in which two involved objects are situated or moving on the same straight line. Most physicists think that relative speed occurs only in 1-D situations. They are wrong: Relative speed occurs not only in 1-D situations but also in 2-D and 3-D situations.

2-D situation is the case in which two involved objects (an object and an observer, for example) are not situated on the same straight line but both are on the same plane. In **Fig. 1-1**, Einstein and the car is in 1-D situation whereas the car and Galileo is in 2-D situation.

3-D situation is the case in which the lines of motion of two objects are not in the same plane (see **Chapter 3** and **Appendix-1**).

Relativists confuse or misidentify relative speed with proper speed. This custom came from Einstein; Einstein took (mistook) proper speed, which is observer-free, for relative speed, which is observer-dependent (see **Chapter 7**). [2] But, how in the world can observer-free speed be relative speed with respect to observers? Einstein did not use the term *proper speed* itself because he thought that all motion is relative and therefore all speed is relative.

Modern relativists use relativistic term ***proper velocity***, which is based on the assumption that distance contraction and time dilation really occur (see **Chapter 7**). [3] The concepts of relative speed and proper speed in this book are mine if not indicated otherwise.

The method of determining the relative speed between two objects is as follows: In 1-D situation, we simply find the difference between the proper speed of one party and the proper speed of the other. For example, in **Fig. 1-1** the proper speed of the car is v and the proper speed of Einstein is zero (0). Therefore, the relative speed between the car and Einstein is $v \sim 0 = v$. If an object is stationary in a given inertial system, the proper speed of the object is zero (0). We cannot use this method in 2-D or 3-D situations. In 2-D or 3-D situations we should use calculus, trigonometry (cosine effect) or speed gun (see **Chapters 4-13**).

The distance between two objects may increase (+) or decrease (−) or remain unchanged with the passage of time. In these cases, the sign of relative speed can be plus (+) or minus (−), or zero (0) respectively. But proper speed is either positive (+) or zero (0); it never becomes negative (−). For example, the proper speed of sound (in a certain atmospheric condition) is always positive (+) regardless of the direction of the propagation of sound.

The proper speed of light in free space (absolute space!) is also constant (c) and positive (+) regardless of its direction or independently of the motion/position of observers (see **Chapters 21-23**). But when it comes to relative speed of light with respect to an observer or any counterpart object increases or decreases or remains unchanged (0). For example, if two photons move in the same direction on the same straight line, the relative speed between the two is $c \sim c = 0$. But according to relativists, who do not discern proper speed from relative speed, $c + c = 0$

and $c - c = c$ (see **Chapter 23**).

Actual Motion/speed and Virtual Motion/speed

Actual motion is *the actual change in the position of an object in a given inertial environment*, and *actual speed* is the rate of actual motion (my definition). *Virtual motion/speed* is *any motion/speed that is not actual in the given environment*. For example, in **Fig. 1-1,** the actual speed of the car is v, the actual speed of Einstein is zero (0), and the actual speed of Galileo is zero (0), respectively.

Relative speed is either *actual speed* or *virtual speed*. For example, in **Fig. 1-1**, the relative speed (v) of the car with respect to Einstein is actual speed, but the relative speed (v) of Einstein with respect to the car is virtual speed because Einstein is not moving actually at this speed in the given environment. The relative speed between the car and Galileo is less then v and changes continuously (see **Chapters 4** and **8**). But the relative speed of the car with respect to Galileo is virtual speed, and the relative speed of Galileo with respect to the car is also virtual speed; neither the car nor Galileo moves actually at such speed in the given environment.

Proper speed is a kind of actual speed. But actual motion/speed is not necessarily uniform and linear. If actual speed is linear and uniform (constant), it is called *proper speed* (my terminology and my definition). The speed appears on the speedometer of a car, for example, is the actual speed of the car in the given environment regardless of whether the car is in linear motion, curvilinear motion, uniform motion, or non-uniform motion. The speed appears on the speedometer of the car is not affected by the motion/position of any third parties (observers or any other objects.) That is, actual speed is observer-free (as is proper speed).

What does the difference between actual motion/speed and virtual motion/speed signify? It signifies a lot: Actual motion/speed has to do with energy; virtual motion/speed does not. Only actual motion can work; virtual motion cannot.

Originally, the idea of length contraction of an object came from the actual motion of the object in *absolute space* allegedly filled with *material ether*. Prior to Einstein, ether-believers thought that apace was filled with material ether and that if an object moves in space filled with ether (this is actual motion in absolute space), the length of the object is reduced or compressed a bit in the direction of the actual motion of the object due to the pressure or resistance of the *material ether*.

So length contraction and time dilation effect were believed to occur in one direction in *actual motion* in absolute space. According to such ether theory, relativistic changes would not occur to the object that is stationary in the absolute space no matter at what relative speed (this is virtual speed) the object may move with respect to its counterpart object in actual motion.

Einstein borrowed the idea and math of Lorentz when he formulated special relativity. Lorentz believed in ether, and he did not discern proper speed from relative speed and actual speed from virtual speed. Therefore, Einstein' relativity is an **ether theory** or **the extension of ether theory**. (Note that general relativity is the extension of special relativity (see **Chapter 40**). [6] But in his 1905 paper of special relativity, Einstein discarded ether because his math does not require the presence of ether. Math does not discern actual motion from virtual motion!

So Einstein and his followers discarded the *controversial* ether. At least ether-believers had the rationale of length contraction and time dilation. But Einstein and his followers have lost such rationale. Theoretically, the mass increase, length contraction, and time dilation of an object and the motion of its counterpart object at a (any) distance have no physical correlation; Many (not all) dissenters believe in ether. But my position is that space is absolute without ether (see **Chapter 18**).

Average Speed and Instantaneous Speed

Average speed is the magnitude of change in position of an object in a certain period of time divided by the time taken.

Instantaneous speed is the rate of change in position at a specific time point ($\lim\limits_{dt \to 0} \frac{ds}{dt}$). If an object is in uniform linear motion, the instantaneous speed is the same as the average speed. But it is not so in the case of non-uniform motion. If we are to find the instantaneous speed of a non-uniformly moving object at a specific time point, we should use calculus, trigonometry or speed detector. The speed appears on a speed detector is the *instantaneous relative speed* between the speed gun and the object the speed gun aims at regardless of the type of motion of the object and the speed gun (observer).

The concept of instantaneous speed was discovered due to the

Special Relativity

invention of calculus. Before calculus was invented in the 1600's (by Sir Isaac Newton and independently by Baron von Leibniz), scientists had not known the concept of instantaneous speed. But even after the calculus was invented, scientists—including Galileo, Newton, Einstein, Harvard physicist and all other physicists—have considered speed only in terms of 1-D situation and not in 2-D or 3-D situation. (This is the reason most modern physicists do not distinguish the concept of proper speed from that of relative speed.)

What is important in relativity is *instantaneous relative speed* rather than average relative speed. Relativistic changes (distance/length contraction, time dilation, and mass increase) are supposed to occur instantly in accordance with the instantaneous relative speed between to objects.

Uniform Motion and Non-uniform Motion

Uniform motion is the case in which the rate of change of position is constant (fixed). **Accelerated motion**, which is an example of not uniform motion, is the case in which the rate of change in the distance is increased or decreases with respect to time. Acceleration itself may be uniform (fixed) or change with respect to time. Relativists assume that acceleration is uniform, and they deal only with **uniform acceleration** in general relativity.

Relativists think that special relativistic effects (distance/length contraction, time dilation, and mass increase) occur only in uniform linear motion and only in 1-D situations and not in non-uniform motion or curvilinear motion. I think differently; relative motion/speed occurs not only in 1-D bur also 2-D and 3-D situation, and if relativity is true, relativistic changes should occur not only in uniform motion but also in any other type of motion in any kind of situation (1-D, 2-D, 3-D). The followings are my argument:

If special relativity is true, relativistic changes occur in accordance with the **instantaneous relative speed** and not with the **average relative speed**. By the way, uniform motion consists of *uniform instantaneous speeds* whereas non-uniform motion consists of *changing instantaneous speeds*. That is, there is no difference between uniform motion and non-uniform motion in the point that both types of motion consist of "instantaneous speeds" Then relativistic changes should occur in any type of motion and in any kind of situation (1-D 2-D or 3-D). That is, relativistic changes should occur not only in uniform linear motion but also in any kind of motion and in any kind of dimensional situation.

But I do not mean that relativistic changes really occur in all types of motion and in all types of situations. Relativistic changes do not occur at all because relativity is not true, and even if relativistic changes really occur, there is no way for scientists to check or measure such changes (see **Chapter 28-31**).

Einstein dealt with only ***uniform acceleration*** (unintentionally though), and he thought that uniform inertial force from uniform acceleration is the same as gravity. Natural gravity, which comes from a body of matter, causes uniform *acceleration* of a free-falling object. But artificial acceleration is not necessarily or always uniform; acceleration itself can be changed (increased or decreased) uniformly or irregularly with the passage of time in reality. Therefore, inertial force, which comes from acceleration (the change in speed or direction of motion), may be uniform or non-uniform, or irregular. So it is wrong to assume that acceleration is always uniform and thus inertial force is also always uniform. Therefore, Einstein's assumption that the inertial force from acceleration is the same as gravity is not true.

According to Einstein, special relativity holds only in uniform motion and that general relativity holds only in the uniformly accelerated motion (see **Chapters 38**). [4] This means that there is *domain discrimination* between special relativity and general relativity. This argument is wrong. The reason is as follows:

Einstein states in his book *Relativity* that the state of being accelerated is the same as the state of being at rest (see **Chapter 34**). [5] I call it Einstein's ***relativity of acceleration***. If this sort of relativity is true, there is no difference between uniform motion and accelerated motion, and therefore, there cannot be domain discrimination between special relativity and general relativity. Einstein did not know that there is fundamental difference between uniform motion and accelerated motion disproved (see **Chapter 34**).

In addition, in reality there is no perfect uniform motion or perfect uniform acceleration in laboratory or on the earth because the earth itself is in complicated rotation motion and revolution motion in space. This means that the earth is not in uniform linear motion. Therefore none of the relativistic experiments or observations done by relativists on the

Special Relativity

earth is valid.

Linear Speed and Curvilinear Speed

Linear (simple form of *rectilinear*) *speed* is the rate of change in the distance in linear motion. Linear motion can be uniform or non-uniform. Even if an object is in uniform liner motion in a given inertial environment, the relative motion of the object with respect to an observer (or any counterpart object) may be uniform or non-uniform or linear or curvilinear depending on the motion or position of the observer. (In this case the various relative motions of the object with respect to the observer are virtual motion.) Actual curvilinear motion/speed is actually not linear. But when it comes to *instantaneous relative speed* between two involved objects is linear in substance regardless of the types of motion of each of the two objects and regardless of the kinds of dimensional situations the two involved objects. Note that the line that connects two involved objects is always *straight* at any moment.

Angular speed (ω) is the rate of change in the angle in circular motion or rotation motion. The *tangential speed* of a *matter point* on a rotating disc is the angular speed (ω) times the distance (r) of the matter point from the center of the rotation motion. Tangential speed is instantaneous and linear in concept. If a disc is in *uniform* rotation motion, the tangential speed of a specific matter point on the disc is uniform. The uniform angular speed of uniform rotation motion or uniform tangential speed is a kind of *proper speed* which is independent of the motion/position of observers.

In his *disc thought experiment*, Einstein took (mistook) the uniform tangential speed of a rod or clock at the rim of the rotating disc for the relative speed of the rod or clock with respect to *any stationary observers* in the inertial environment/system in which the disc rotates (see **Chapters 13** and **40**). [7] Einstein was wrong: If an observer is at rest in an inertial environment, the relative speed of a matter point on the rim of a rotating disc with respect to the observer is not uniform but changes periodically (see **Chapter 13**).

Absolute Motion/Speed

Absolute motion is the motion of an object with respect to absolute space (more on absolute space and absolute motion in **Chapter 18-22**). Absolute speed is *the actual speed of an object with respect to absolute space*. Absolute motion/speed is independent of observers. Absolute motion is not necessarily uniform; it can be of any type. I suggested a method of detecting the absolute motion/speed of an object with respect to absolute space in **Chapter 22**.

Throughout this book I proved that space and time are absolute entities and that the speed of light is constant with respect to absolute space and not to human observers or objects whose speed and direction are arbitrary and unpredictable. ♦

Special Relativity

2

Reference Body and Coordinate System

The motion or Speed of an object is defined with respect to a specific reference body. There are two kinds of reference bodies—*environmental reference body* (simply *environment*) and ***point reference body*** (my terminology). I use the term *environment* as the synonym of *system* or *coordinate system*.

There are three kinds of environmental reference bodies--a line (straight line), plane, and a body of space, in each of which an object can move *actually* in various ways.

- A line, a one-dimensional environment, has only one axis—x. On a line, an object can move in two ways--back and forward or stay put at a given point.
- A plane--a two-dimensional environment, has two axes—x and y. On a plane an object can move back and forward or up and down or obliquely in various angles.
- A body of space, a three-dimensional environment, has three axes—x, y, and z. In a body of space, an object can move in three-dimensional way.

I do not recognize four or higher-dimensional environment.

A point, observer, observation point, or any object in general, which can be represented as a point, is a ***point reference body*** from which the ***relative*** motion or *speed* of a counterpart point/object is defined.

Relativists confuse or do not distinguish environmental reference body from point reference body. Therefore the concept of relative motion/speed of relativists is vague and incorrect. We need to note the

following facts:
- **Proper speed** or **actual speed** is defined with respect to an environmental reference body whereas **relative speed** is defined with respect to a point reference body.
- A point reference body has no axis. So the relative speed of an object with respect to a point reference body has only the magnitude of relative motion; the direction of the relative motion is not defined. So the term *velocity* (speed + direction) is not proper for relative motion.

We do not define the relative speed of an object with respect to an environmental reference body such as a straight line, plane, or body of space. For example, we do not (cannot) define the relative motion/speed of an object with respect to a straight line or plane or a body of space.

Relativists might think, as Einstein actually did, that we can attach three perpendicular planes to a (any) point or observer or object so that they can define the direction of motion or the coordinate values of a target objet with respect to the point reference body (or observer or object). Relativists are wrong: If we attach three perpendicular planes to a point (or point reference body), it is now a three-dimensional (environmental) reference body.

Reciprocity holds only in relative motion/speed. Reciprocity does not hold in proper motion or actual motion. For example, if a car is in actual speed of v on a straight road or plane; we do not and cannot say that the car is stationary whereas as the straight road or the plane moves at the speed of v with respect to the car. What is the difference? The car uses energy (fuel) to move in the given inertial reference body whereas the road or the plane does not. This is the difference between actual motion and virtual motion. So reciprocity or relativity does not hold in this case.

Relativists might think that if a top rotates in space, it is the same as the situation that the top is stationary and the whole universe is rotating with respect to the top. Relativists are wrong; reciprocity or relativity does not hold in this case either. The rotation of the top is actual motion and the rotation of the universe with respect to the top is virtual motion; centrifugal force (a kind of inertial force) occurs only in actual rotation motion of the top and not in the *virtual rotation motion* of the universe.

A reference body, whether that is an environment or point, is supposed to be in inertial state. Strictly speaking, any object or laboratory on the surface of the earth is cannot be an ***inertial reference body*** because the earth itself is not in inertial state. This means that relativistic observations or experiments done on the surface of the earth are not valid; the results or data from such observations or experiments fall in the range of experimental errors and thus do not support relativity. ♦

3

Kinds of Dimensional Situations

Relativists think of relative speed in terms of 1-D (one-dimensional) situation. However, relative speed occurs not only in 1-D situations but also in 2-D or 3-D situations. We need to understand three dimensional situations (1-D, 2-D, and 3-D situations) to better understand the correct meaning of relative speed.

1-D situation is the case in which two objects are situated or moving on the same straight line. **Fig. 4-1** shows two examples of 1-D situation.

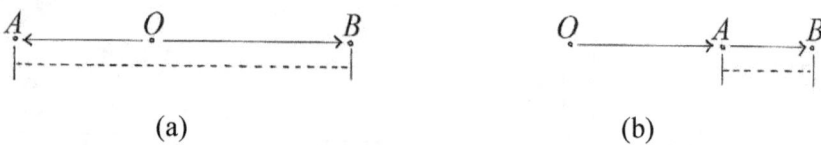

Fig. 4-1 Two examples of 1-D situation

In diagram (a) in above diagram, two objects A and B started from common origin (or observer) O at the same time and have moved for one second in the opposite directions of each other. Observer O is stationary in the given inertial environment (straight line). In this case, the following are true.

- Distance OA represents the magnitude of the *proper speed* of A.
- Distance OB represents the magnitude of the *proper speed* of B.
- Distance AB (dotted line) represents the magnitude of the *relative speed* between A and B.

In diagram (b) in **Fig. 4-1,** two objects A and B started from

Special Relativity

common origin (or observer) O at the same time and have moved for one second in the same direction. Observer O is stationary in the given inertial environment (straight line). In this case, the following are true:

- Distance OA represents the magnitude of the *proper speed* of A.
- Distance OB represents the magnitude of the *proper speed* of B.
- Distance of AB (dotted line) represents the magnitude of the *relative speed* between A and B.

2-D situation is the case in which two involved objects are not situated on the same straight line yet both are on the same plane. **Fig. 4-2** shows three examples of 2-D situation.

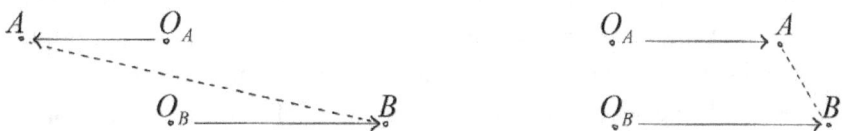

(c) Two lines of motion are parallel.   (d) Two lines of motion are parallel.

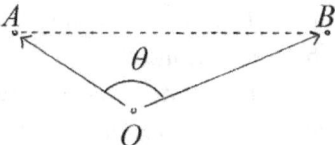

(e) Two lines of motion meet or intersect.
Fig. 4-2 Three examples of 2-D situation.

In diagrams (c) above, two objects A and B started from origins O_A and O_B respectively at the same time and have moved for one second in the directions indicated by arrows. The line of motion of A and that of B are parallel to each other. In this case, the following are true:

- Distance $O_A A$ represents the magnitude of the *proper speed* of A.
- Distance $O_B B$ represents the magnitude of the *proper speed* of B.
- Distance AB (dotted line) **does not** represent the relative speed between A and B. We need to use calculus or radar to find the relative speed between A and B. See the case of parallel motion in **Chapter 11**.

In diagrams (d) in **Fig. 4-2**, two objects A and B started at the same time from two origins O_A and O_B respectively and have moved for one second in the directions indicated by arrows. The line of motion of A and that of B are parallel to each other. In this case, the following are true:

- Distance $O_A A$ represents the magnitude of the *proper speed* of A.
- Distance $O_B B$ represents the magnitude of the *proper speed* of B.
- Distance AB (dotted line) ***does not*** represent the relative speed between A and B. The rate of change of distance AB is the relative speed between A and B. We should use calculus or radar to find the relative speed between A and B. See the case of parallel motion illustrated in **Exercise A1-1** in **Appendix-1**.

In diagram (e) of **Fig. 4-2**, two objects A and B started from common origin O at the same time and have moved for one second maintaining a certain intersection angle θ (AOB). In his case the following are true:

- Distance OA represents the magnitude of the *proper speed* of A.
- Distance OB represents the magnitude of the *proper speed* of B.
- Distance AB (dotted line) ***does not*** represent the relative speed between A and B. The rate of change of distance AB is the relative speed between A and B. We need to use trigonometry and calculus to find the relative speed between A and B. See the example illustrated in **Chapter 10**.

3-D situation is the case in which two lines of motion of two objects are not in the same plane. In other words, the two lines of motion do not meet (intersect) each other. Diagram (f) in **Fig. 4-3** is an example of 3-D situation.

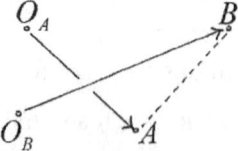

(f) Two lines of motion are not in the same plane.
Fig. 4-3 An example of 3-D situation.

In diagram (f) in above diagram, two objects A and B started from origins O_A and O_B respectively at the same time and have moved for one

Special Relativity

second in the directions indicated by arrows. In this case, the following are true:

- Distance $O_A A$ represents the magnitude of the *proper speed* of *A*.
- Distance OBB represents the magnitude of the proper speed of B.
- The rate of change of distance *AB* (dotted line) is the relative speed between *A* and *B*. We need to use trigonometry and calculus to find the relative speed between *A* and *B*. See the examples in **Exercises A1-5** and **A1-6** in **Appendix**.

So far we have discussed the concept of relative speed in three kinds of dimensional situations—1-D, 2-D and 3-D situations in each of which the involved objects are in *uniform motion*. The case in which a target object is not in *uniform linear motion* is more complicated (see **Chapters 12 ~14)**. ♦

4

The Case in Which an Observer Sits Off the Line of Motion of a Uniformly-moving Object

In the diagram below, object A is in uniform motion along straight line EF at the speed of v in the direction $E \to F$. E and F are points that are very far from each other. Observer O is distance r from EF. H is the point at which the perpendicular from O to EF meets EF. That is, $OH = r$. Observer O is stationary in the given inertial system.

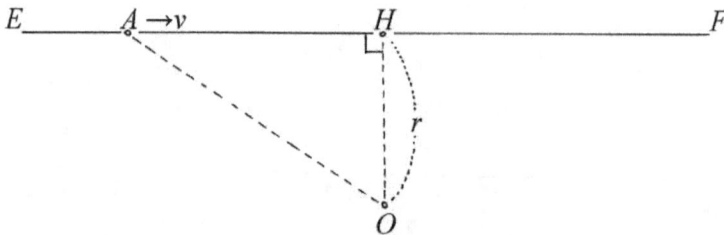

Fig. 4-1 Observer O is distance r from EF.

Let us solve the two problems as follows:

(1) Prove or disprove that the relative speed of A with respect to O is v and constant.

(2) Find the relative speed of A with respect to O at the moment A passes point H.

Practically all physicists (both relativists and dissenters) I have contacted so far (2001-2013) think that the relative speed of object A with respect to observer O is v *and constant*. And they think that the

Special Relativity

relative speed of A with respect to O at the moment A passes point H is v and constant. They are wrong.

The given situation in **Fig. 4-1** is 2-D. We can solve the given problems by finding the equation that represents distance AO and then another equation that represents the relative speed of A with respect to or vice versa.

In **Fig. 4-2** below, object A moves from E to F at a uniform speed v. Point A_0 is distance m from H. Let us assume that the time A passes point A_0 is $t = 0$ (zero). Then, $A_0 A = vt$.

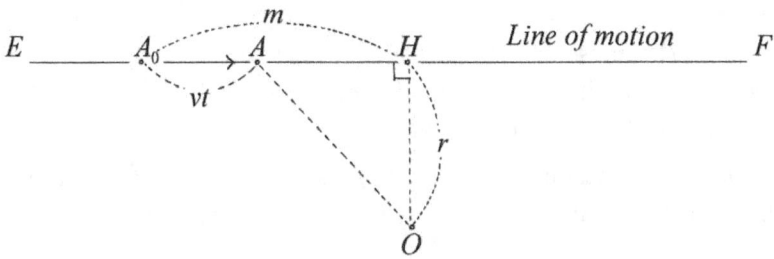

Fig. 4-2 The time A passes point A_0 is $t = 0$.

Let us distance AO be Y and find the equation that represents Y. AOH is a right triangle. $AH = |m - vt|$, and $OH = r$. Therefore,

$$AO = Y = \sqrt{(AH)^2 + (OH)^2}$$
$$= \sqrt{(vt - m)^2 + r^2}. \quad (1)$$

Eq. (1) represents distance AO at time t.

The derivative (Y') of **Eq. (1)** with respect to time t is

$$Y' = \frac{dy}{dt} = \frac{v(vt - m)}{\sqrt{(vt - m)^2 + r^2}}. \quad (2)$$

Eq. (2) represents the relative speed of A with respect to O (or vice versa) at any given time t. **Eq. (2)** shows that the relative speed of A with respect to O varies continuously as time t changes (lapses). That is, the relative speed between A and O is neither v nor constant.

Now, let us find the relative speed between A and O at the time A

passes point *H* by using **Eq. (2)**.

The time *A* passes point *H* is $t = m/v$. By applying this time value ($t = m/v$) to **Eq. (2)** we find

$$Y' = \frac{v(m-m)}{\sqrt{(m-m)^2 + r^2}} = \frac{0}{r} = 0.$$

This means that the relative speed of *A* with respect to *O* (or vice versa) at the moment *A* passes point *H* is zero (0)! (Note that the *proper speed* of *A* is still *v* at this moment.) Quite an unexpected result!

I have explained this result to many physicists. But none of them understood my explanation. Relativists and dissenters did not understand why **Eq. (2)** represents the relative speed between *A* and *O* in the first place. Physicists think that relative speed exists only when two objects are situated in 1-D situation.

Generally speaking, if $Y [= f(x)]$ is the equation (function) that represents the distance between two objects with respect to time *t*, the derivative ($\frac{dy}{dt}$) of the equation represents the *rate of change in the distance* (= instantaneous relative speed or simply relative speed) between the two at time *t*. This is only a high school level of math.

How do we know whether **Eq. (2)** is correct?

We can prove **Eq. (2)** by replacing *r* with zero (0). (If $r = 0$, this is a 1-D situation.)

If $r = 0$ (zero) in **Eq. (2)**,

$$Y' = \frac{v(vt-m)}{\sqrt{(vt-m)^2 + r^2}} = \frac{v(vt-m)}{\sqrt{(vt-m)^2}} = v.$$

(This is when observer *O* sits on the line of motion of *A*.)

Therefore, we can see that **Eq. (2)** is correct.

Now, I would explain why the relative speed between *A* and *O* in **Fig. 4-2** becomes zero (0) at the moment *A* passes point *H*.

During the time *A* moves from *E* to *H* (see **Fig. 4-2**), distance *AO* *decreases* (−) continuously with the lapse of time. (The rate of decrease of distance *AO* is not constant; it varies! See the speed graphs in **Fig. 4-3**).

Special Relativity

On the other hand, during the time A moves from H to F, distance AO **increases** (+) continuously with the passage of time. (The rate of increase of distance AO varies; see the speed graphs in **Fig. 4-3**). Then, what about the moment A passes point H? At this moment (this is when $t = m/v$) distance AO neither decreases (−) nor increases (+).

When a value changes continuously from negative value to positive value, it surely passed the moment at which the value becomes zero (0). H (see **Fig. 4-2**) is the point at which the rate of change in the distance AO turns from decrease (−) to increase (+). That is, distance AO neither increases nor decreases at this moment.

What is the relative speed between A and O at the moment distance AO does not change? The answer is zero (0). (Yet the *proper speed* of A on its line of motion at this moment is v.)

I refer to angle AHO as **observation angle** or **angle of observation** (θ), Observation angle means the angle formed by two lines—the line of motion of an object and the line that connects the object with an observer. Note that when $t = m/v$, the observation angle (θ) is 90°. If $r = 0$, observation angle becomes either 0° or 180° (see more about *observation angle* in **Chapter 8**).

If $m = 0$ in **Fig. 4-2**, **Eq. (2)** becomes as follows:

$$Y' = \frac{v(vt)}{\sqrt{(vt)^2 + r^2}}. \tag{3}$$

Eq. (3) means that the time A passes point H is zero ($t = 0$). What is the relative speed of A with respect to O at the moment A passes point H in **Eq. (3)**?

By applying $t = 0$ to **Eq. (3)** we find

$$Y' = \frac{0}{\sqrt{(0)^2 + r^2}} = 0.$$

Both **Eqs. (2)** and **(3)** show that the **denominator** is always larger than the numerator due to the r factor in the denominator. This means that the relative speed of A with respect to O is always less than the proper speed (v) of A and changes continuously at *varying rates*.

Fig. 4-3 shows the relative speed graphs of **Eqs. (2)** and **(3)**. Graph (a) is of **Eq. (2)**, and graph (b) is of **Eq. (3)**.

If $t = 0$ in **Eq. (2)** [see graph (a) in **Fig. 4-3**], the relative speed between A and O is

$$Y' = \frac{v(-m)}{\sqrt{(-m)^2 + r^2}} = \frac{-vm}{\sqrt{m^2 + r^2}}.$$

If $t = 0$ in **Eq. (3)** [see graph (b) in **Fig. 4-3**], the relative speed between A and O is

$$Y' = \frac{v(vt)}{\sqrt{(vt)^2 + r^2}} = 0.$$

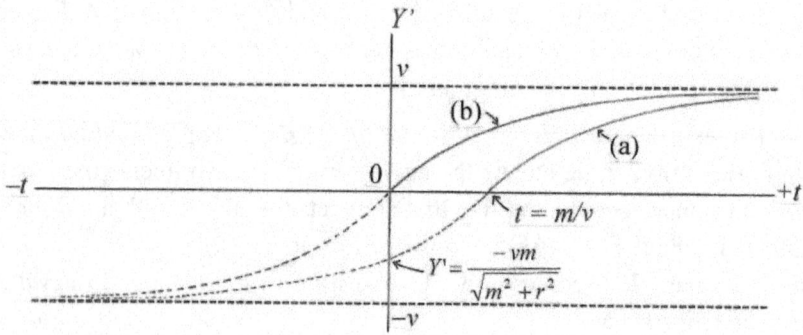

Fig. 4-3 Graph (a) is of **Eq. (2)**. Graph (b) is of **Eq. (3)**.

Both graphs (a) and (b) are the same except their positions on the t (time) axis. This difference comes from the factor m (see **Fig. 4-2**). Both graphs (a) and (b) show that the relative speed of A with respect to observer O changes continuously with the change of time (t). This is so in spite of the fact that A is in uniform motion and observer O is stationary in the given frame.

Here is the summary of **Fig. 4-3**:

If $r = 0$ in **Eq. (2)** or **Eq. (3)**, (now the situation is 1-D), $Y' = v$. This means that v_{AO}—the relative speed of A with respect to O—equals the proper speed (v) of A.

If $r \neq 0$, (now is a 2-D situation), v_{AO} changes continuously within the range of $-v < v_{AO} < v$.

If observation angle θ becomes 90°, $v_{AO} = 0$ (zero). This is so even when the object or observer or both are in uniform motion, accelerated

Special Relativity

motion, rotation motion, or irregular motion (see **Chapters 5, 8, 9, 10, 11, 12,** and **13**).

Observation angle of 90° is formed in 2-D situations. But observation angle 90° can be formed in 3-D situations (see **Exercises A1-5 and A1-6** in **Appendix A-1**).

I refer to the above results (facts) as the first law, the second law, and the third law of relative speed respectively as follows:

The first law of relative speed

In a 1-D situation, the relative speed between two involved objects is the difference between the proper speeds (or actual speeds) of the two objects ($= |v_1 - v_2|$). (See the definition of actual speed in **Chapter 1**.)

The Second law of relative speed

In a 2-D or 3-D situation, the relative speed between two involved objects is not simply the difference between the actual speeds of the two objects ($\neq v_1 \sim v_2$). In this case we can find the relative speed between the two objects by using calculus or trigonometry (cosine effect) or radar.

The third law of relative speed

At the moment the *observation angle* is 90°, the relative speed between two involved objects is zero (0) regardless of the types of motion or magnitudes of the actual speeds of the two objects. (See *the fourth law of relative speed* in **Chapter 6**.)

The three laws of speed above are applied even to the speed of light. For example, if we substitute object *A* in **Fig. 4-1** (or **Fig. 4-2**) with a photon, the relative speed of the photon with respect to observer *O*, who is off the line of motion of the photon, is less than c and changes continuously as the photon moves. And if the observation angle becomes 90 degrees, the relative speed of the photon with respect to observer *O* becomes zero (0)! We will discuss the speed of light more in **Chapters 17-27**.

Physicists and Relative Speed

In 1994 or so, while I was walking in the garden of Kyung Bok Kung Palace (Seoul, South Korea) I vaguely thought that if an observer sits off the line of motion of a uniformly-moving object (this is a 2-D situation as shown in **Fig. 4-1**), the relative speed of the object relative to the observer would not be constant. But I did not give further thought to this idea because I thought that this was quite a complicated problem. In 2000 winter, I began to study relative speed in a 2-D situation using calculus, and I found **Eq. (3)** (see p. 73). This was a ***Eureka*** moment; I was excited. At that time I was a sophomore student studying art/painting at the Southern Connecticut State University (SCSU).

At the beginning of the spring semester of 2001, I asked a leave of absence for a semester with the excuse that I would write a science paper. I wrote 136 pages-long book (manuscript) titled *Relativity, a Mistake of the 20th Century*. I registered this book in 2001 but had not published it because responds from physicists whom I contacted were unanimously negative or silent. I returned to school that fall semester that year (2001). I remember that my depression was doubled with the 9-11 attacks at the beginning of the fall semester.

After I graduated magna cum laude from the SCSU in 2002, I began to improve my physics paper. In 2005, I attended the 12th NPA (Natural Philosophy Alliance) Conference (May 23 –May 27, 2005) at Storrs, Connecticut with my paper entitled "The Speed of an Object with Respect to an Observer Who is off the Line of Motion of the Object." [1] The NPA is an organization of physicists who do not agree with relativity. At the conference, I explained the relative speed in a 2-D situation to physicists. But not a single physicist understood what I said. After the NPA conference, I began to study further the method of finding relative speeds in various cases of 2-D and 3-D situation.

In 2006, I began to send to my improved papers famous physicists of America and the world in the hope they might understand the concept of relative speed in 2-D and 3-D situations.

Dr. Roy J. Glauber and many other physicists of the Harvard University, Dr. Steven Weinberg of the University of Texas at Austin, and many other physicists did not pay attention to my paper. Physicists (both relativists and dissidents) just do not (want to) know what relative speed is; they seem to be unable to understand the meaning and use of differential calculus in relation with relative speed. The reason is that physicists' concept of relative speed is limited to 1-D. Calculus is an indispensable tool in understanding and finding the relative speed in 1-D, 2-D, and 3-D situations.

Special Relativity

HARVARD UNIVERSITY
DEPARTMENT OF PHYSICS
17 OXFORD STREET
CAMBRIDGE, MA 02138

ROY J. GLAUBER
Mallinckrodt Professor of Physics

December 6, 2006

Mr. Byoung Ha Ahn

Dear Mr. Ahn,

    Several days ago you sent me a manuscript on motions in the special theory of relativity. I want to thank you for sending me your thoughts, but have to tell you that I have not had the time to read your manuscript. The calculations you have done are evidently rather time-consuming.

    I want to thank you also for the thoughtfulness of the $100 check you sent me. I am returning it herewith, as I don't think I have really been able to help you.

Sincerely yours,

Roy J. Glauber

Fig. 4-4 The answer letter I got from Dr. Glauber. I had included a $100 check to my paper as a reviewing fee. I sent copies of my letter and relativity essay to many physicists. Dr. Glauber was the only person who bothered to respond to my then 45-page paper. [2] I blocked the address or email address of the sender and the receiver in the above letter. ♦

5

r-factor

r- factor (or *r factor* or simply *r*) is the distance of an object from the line of motion of a counterpart object (my terminology and my definition). The term *line of motion* means a straight line (when an object is in linear motion) or tangent line (when an object is in curvilinear motion) (my terminology and my definition).

We have dealt with the concepts of *r*-factor and the line of motion in the previous chapter. *r*-factor is found in the denominators of **Eqs. (2)** and **(3)** (see **Chapter 4**). Therefore, *r*-factor inversely affects the relative speed of a target object with respect to an observer. In other words, the larger the value of *r* (= the farther the distance of an observer from the line of motion of target object) the smaller the absolute value of the relative speed of the target object with respect to the observer.

What if *r* becomes infinite (∞)?

If $r = \infty$ in **Eq. (2)**,

$$\lim_{r \to \infty} \frac{v(vt-m)}{\sqrt{(vt-m)^2 + r^2}} = 0.$$

If $r = \infty$ in **Eq. (3)**,

$$\lim_{r \to \infty} \frac{v^2 t}{\sqrt{(vt)^2 + r^2}} = 0. \quad \text{[This is when } m = 0 \text{ in } \textbf{Eq. (2)}.\text{]}$$

The above results means that if an observer is very far (∞) from the line of motion of a target object (see **Fig. 4-2**), the relative speed of

Special Relativity

the object with respect to the observer becomes zero (0) even when the observation angle (θ) is not 90° *regardless* of the magnitude of the proper speed (*v*) of the object on its line of motion.

Fig. 5-1 shows five graphs of **Eq. (3)** with five different values of *r*. Graph (a) is when $r = 0$ (this is a 1-D situation). Graph (e) is when $r = \infty$. The order of values of *r* in graphs (a), (b), (c), (d), and (e) are $r_a < r_b < r_c < r_d < r_e$. These graphs show that bigger the value of *r*, the flatter and closer the graph becomes to the *t*-axis.

If $r = \infty$, the speed curve (e) becomes a straight line and coincides with *t*-axis. That is, the relative speed between an observer *0* and object *A* (see **Fig. 4-1** or **Fig. 4-2**) is zero (0) regardless of the passage of time (*t*).

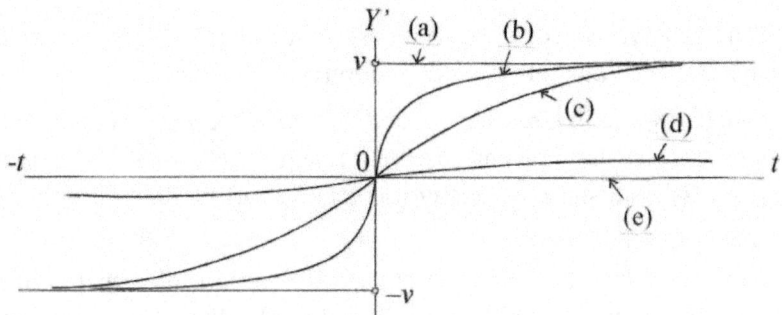

Fig. 5-1 The effects of factor *r* in **Eq. (3)**

Graph (a) in **Fig. 5-1** calls for our special attention: graph (a) is the case when $r = 0$ in **Eq. (3)** (see **Fig. 4-1**). The other four graphs [(b), (c), (d), and (e)] are all continuous lines; in each of these four cases, the relative speed *v* is zero at $t = 0$. But in graph (a), the line is broken at $t = 0$ and the relative speed is neither *v* nor *–v* nor zero (0) at this moment [note the two hollow dots in graph (a) at the moment $t = 0$]. Let us see why: This is the case in which observer *O* sits at point *H* (see **Fig. 5-2**). (Note that *O* and *H* coincide.) In **Fig. 5-2**, object *A* moves from *E* to *F* at the *uniform speed v*. *EF* is a straight line. So this is a 1-D situation.

Fig. 5-2. Observer *O* is at point *H*. The time *A* passes *O* is $t = 0$.

During the time object *A* (see **Fig. 5-2**) moves from *E* to *O*, distance *AO* decreases (−) at the uniform rate *v*. Therefore, the relative speed of *A* with respect to *O* is − *v* in this time period [see the portion of graph (a) where $t < 0$ in in **Fig. 5-1**.]

On the other hand, during the time *A* moves from point *O* to *F* (see **Fig. 5-2**), distance *AO* **increases** (+) at the uniform rate *v*. Therefore, the relative speed of *A* with respect to *O* in this time period is +*v* [see the portion of graph (a) where $t > 0$ in **Fig. 5-1**].

Then, what is the relative speed of *A* with respect to *O* at the moment $t = 0$ (this is when *A* passes point *O*)? This is the moment the relative speed of *A* with respect to *O* changes abruptly (not gradually) from −*v* to +*v*. My answer for this question is as follows:

The relative speed of *A* with respect to *O* does not exist at the moment *A* meets (hits) observer *O* (= observer).

We have seen in **Chapter 4** that the relative speed between *A* and *O* becomes zero at the moment observation angle *AHO* is 90° (this is the story in a 2-D situation in which observer *O* is off the line of motion of *A*).

However, **Fig. 5-2** is a 1-D situation; therefore observation angle *AHO* is not formed at the moment *A* passes *O*. (At this moment, three points *A*, *H*, and *O* meet at the same point.) At his moment, observation angle *AHO* is neither 90 degrees nor 0 degrees; no observation angle and no relative speed at this moment.

Jokingly speaking, this is the moment observer *O* is hit by object *A* and thus the observer loses his/her consciousness. So the observer forgets everything about relative speed, proper speed, Newton, Einstein, relativity, and everything else at this moment.

> In 3-D situations, *r*-factor is redefined as "the distance between the line of motion of an object and the line of motion of the counterpart object" (see **Exercises A1-5** and **A1-6** in **Appendix-1**). ♦

Special Relativity

6

The Secrets of Fixed Stars

Fixed stars look stationary. This phenomenon has to do with the large (∞) magnitude of *r*-factor which we have discussed in the previous chapter (**Chapter 5**). Let us see why fixed stars look stationary in the following exercise:

Exercise 6-1

In the diagram below, heavenly body A moves at a uniform speed of 0.8 c (= 240,000km/sec) in its line of motion *EF*. Observer O is one hundred million light years away from *EF*. H is the point at which the perpendicular from O to *EF* meets *EF*. That is, OH (= r) = 100,000,000 light yrs. A_1 is the position of heavenly body A at the moment A has moved for one second from H toward F.

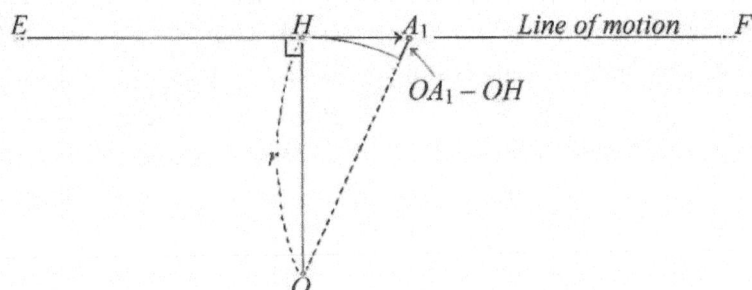

Fig. 6-1 r is very large (∞).

Find the followings:

1. The *average* speed of A relative to O during the time A moved

Harvard Physicists Confuse Relative Speed with Proper Speed

from H to A_1.

2. The speed of A relative to O at the moment A passes A_1.

3. One light year is 9.46 trillion km. The speed of light (c) is 300,000 km/sec. Assume that observer O is stationary in the given system.

Solution:

The change of distance AO during the given one second is not HA_1 but "$OA_1 - OH$". (Almost all physicists do not understand this!) Therefore, the *average* relative speed of heavenly body A with respect to observer O during this time (one second) is

$$(OA_1 - OH)/sec = \left\{\sqrt{(OH)^2 + (HA_1)^2} - OH\right\}/sec$$

(Since $OH = 9.46 \times 10^{20}$ km and $HA_1 = 240{,}000$ km,)

$= 0.000{,}000{,}000{,}030$ km/sec (or $0.000{,}0030$ cm/sec or 0.030 μ/sec).

(See the manual calculation in **Fig. 6-3** at the end of this chapter.)

However, the instantaneous relative speed of A with respect to O at the moment A passes point A_1 is 0.064 μ/sec, and the speed of A with respect to O at the moment A passes point H is $0 km/sec$. These values (0.064 μ/sec and 0 km/sec) are obtained by using **Eq. (3)** (see **Chapter 4**).

The relative speed of the star ($0.030\mu/sec$ or $0.064\mu/sec$) with respect to observer O is negligibly small compared to the proper speed (240,000 km/sec) of heavenly body A in its line of motion. These unexpected and unbelievable results are due to the large r factor (=the distance between observer O and the line of motion of heavenly body A). $r = 100{,}000{,}000$ light years in the above case is still not that large on the universal scale.

What if r is truly infinite (∞)? If $r = \infty$, the relative speed of heavenly body A with respect to observer O becomes 0 (zero) even when the *observation angle* (θ) is not 90°. This is true even when the proper speed of A is equal to or faster than the speed of light (c).

$$\lim_{r \to \infty} \frac{c^2 t}{\sqrt{(ct)^2 + r^2}} = 0.$$

(This is so as long as t is not large enough to cancel the effect of $r = \infty$.)

I refer to this phenomenon as **the fourth law of relative speed**.

Special Relativity

> **The Fourth Law of Relative Speed**
> If *r*-factor is very large (∞), the relative speed between two involved objects is zero (0) even when the observation angle is not 90°.

Eq. (3), from which we have just derived the fourth law of relative speed, is the case of 2-D situation. But the fourth law holds not only in 2-D situations but also in 3-D situations (see **Exercises 1-5** and **1-6** in the **Appendix-1**). In reality, the earth and a (any) fixed star are normally in a 3-D situation.

Fixed stars are those whose distances from the earth are very large. Fixed stars look stationary in the sky. But actually, fixed stars are moving at astronomical speed in various different directions. But their relative speeds with respect to the earth are very small because of *the fourth law of relative speed*. *The World Book Encyclopedia* states about fixed stars as follows: [1]

> Fixed star is an expression often used in referring to the stars, because their places in the sky relative to one another do not seem to change. Actually, however, the stars are moving in many directions, and the pattern of the heavens is slowly changing.
>
> But the changes are scarcely noticeable within a person's lifetime because those stars are so far away and the distances between them are so great. Even Barnard's star, the one believed to move the fastest, changes position by a distance equal only to the moon's diameter in 200 years. Compared to the planets, which can be seen constantly shifting their positions in the sky, the starry background seems fixed. Astronomers use photography to study the motions of all the bright stars and many faint ones. In photographs taken at different times many years apart, they compare the positions of the stars and note how they have changed.

The North Star (Polaris) is approximately aligned with the direction of the rotational axis of the earth. Therefore, the North Star seems to stay put at a point in a photo taken with the shutter open at night in the north hemisphere (see **Fig. 6-2** on next page).

The North Star cannot be observed in the southern hemisphere. Photographs show us that the position of the North Star is quite fixed at a point. But other fixed stars seem to revolve round the North Sate once a day drawing big circle when seen in the photos taken in the northern hemisphere. Does this phenomenon contradict with the argument that the relative speed of any (each) fixed star relative to the earth is zero (0)? Not really. Here is the explanation:

The diameter of the earth or the solar system or even the galaxy system, to which the earth belongs, is only a *point* (mathematical point) if viewed from a fixed star who is really infinite (∞) distance from our Galaxy. The daily circular motion of a (any) fixed star is a ***virtual motion*** and not ***actual motion***; Fixed stars do not actually revolve around the earth; the circular motion of the fixed stars around the earth is virtual motion due to the rotation of the earth. Since the earth (and even our Galaxy) is only a mathematical point when viewed from the fixed star, earth' rotation and revolution motion within the Galaxy does not affect the relative speed between the star and the earth. Therefore, the relative motion/speed of the earth with respect to any fixed stars is zero (0). (This is so as long as the earth is not aligned with the direction of motion of the fixed star.)

Fig. 6-2 A long exposure photo of Polaris and neighboring stars (exposure time 45 min).
(Source: http://en.wikipedia.org/wiki/Pole_star)

Special Relativity

Some readers might think we may be able to set the ***absolute coordinate system*** so that we can give absolute coordinate values to any heavenly body at a specific time point by making the most of fixed stars. But the author thinks we cannot. Most fixed stars are not perfectly fixed. In addition, the author proved *mathematically* that the relative speed of the earth (or any object) with respect to a (any) fixed star is ***zero*** (0) or almost zero in spite of the absolute motion of the earth in space. Therefore, we cannot give the absolute coordinate values to a point (object) in space with respect to fixed stars. But the fact that we cannot give specific coordinate values to a point in space does not disrupt the absoluteness of space and time.

Some relativists might think that since most heavenly bodies (including our earth) rotate, we cannot find the relative speed between heavenly bodies. Einstein states in his book *Relativity* that an object should be in ***translatory motion*** and not in rotatory motion if we are to find the relative speed of the object (see **Chapter 7**). [2] Einstein is wrong; relative speed occurs/exists even if an object or an observer or both are in any kind of motion.

Translatory Motion and Rotatory Motion

Source:
http://in.answers.yahoo.com/question/index?qid=20071206134641AAl2bWb

The best way to decide if the body is making a purely translatory motion or not is as under.

Join any two points of a body by a straight line. Irrespective of the selection of the points if during the motion, the line moves parallel to itself, the body has purely translatory motion. Such a motion may be along a straight line or on a curvilinear path.

If a body is moving on a circular path such that the straight line joining any two points of it moves parallel to itself, it is a purely translatory motion. If all points on it move on circular path with a common centre, then it is a purely rotational motion. If none of these two conditions is satisfied, then it is a mixture of translatory and rotatory motion.

Harvard Physicists Confuse Relative Speed with Proper Speed

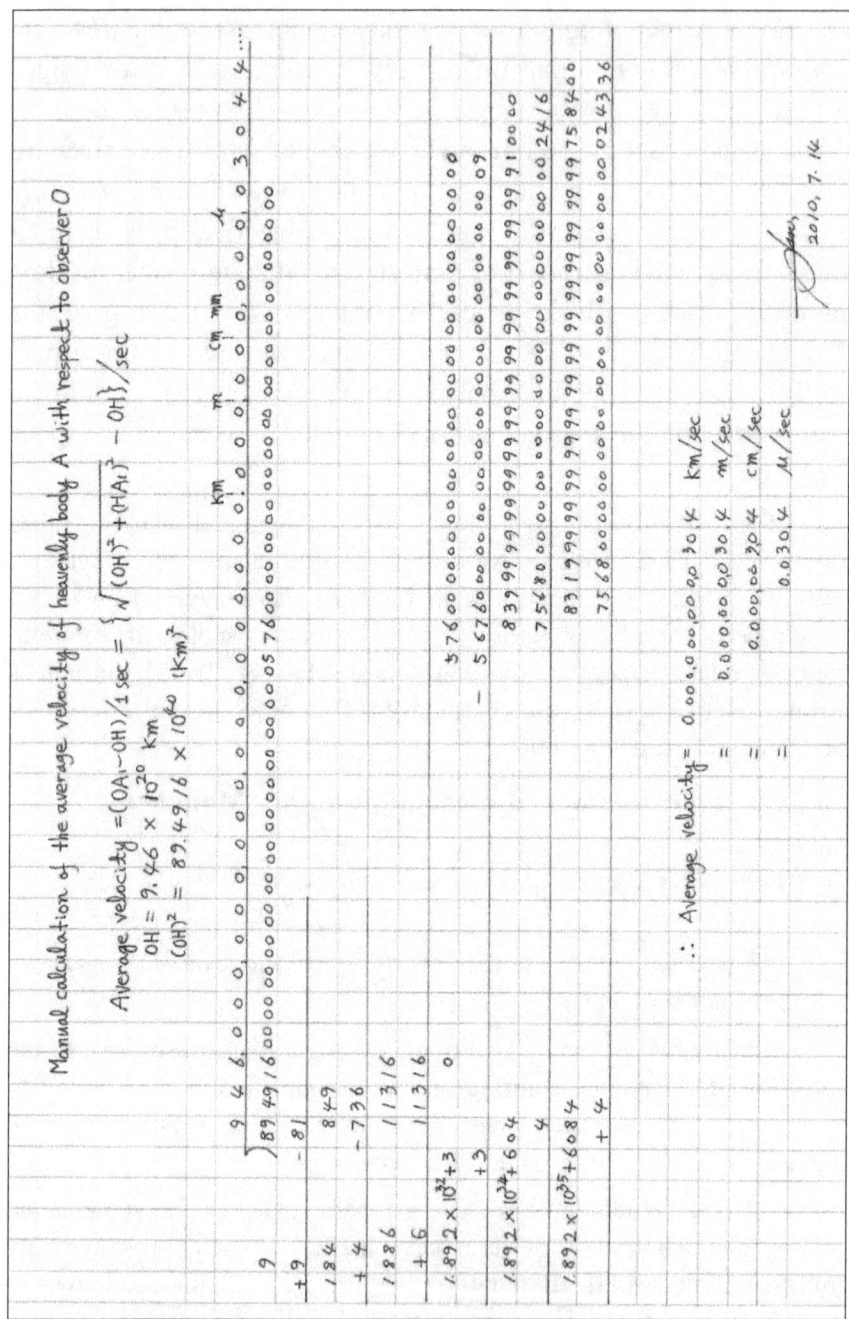

Fig. 6-3 Manual Calculation of **Exercise 6-1**.
(Note: I made the calculation in 2006, revised it in 2007, and redid it in 2010 for a print version. ♦)

7

Einstein Mistook Proper Speed for Relative Speed

Einstein thought that all motion/speed is relative. Einstein did not distinguish *proper velocity/speed* (which is observer free) from relative speed (which is observer-dependent). (Note: **Chapters 1-6** are prerequisite for this chapter.)

Let us see how Einstein mistook proper speed for relative speed in his *raven thought experiment* in his book *Relativity*: [1] (The underlines and the numbers given to underlined parts are mine.)

> Let us imagine a raven flying through the air in such a manner that its motion, as observed from the embankment, is uniform and in a straight line. (#1) If we were to observe the flying raven from the moving railway carriage, we should find that the motion of the raven would be one of different speed and direction, but that it would still be uniform and in a straight line. (#2) Expressed in an abstract manner we may say: If a mass m is moving uniformly in a straight line with respect to a co-ordinate system K, then it will be moving uniformly and in a straight line relative to a second co-ordinate system K', provided that the latter is executing a uniform translatory motion with respect to K.

Fig. 7-1 (on next page) is the situation which Einstein means in his raven thought experiment. In the diagram, the proper velocity (speed+ direction) of the raven with respect to the first inertial environment, which includes the embankment and observer O_1, is constant ($= v_1$) (see underlined part #1 in the above citation) independently of the motion/position of observer O_1 or other observers (not shown) on the embankment.

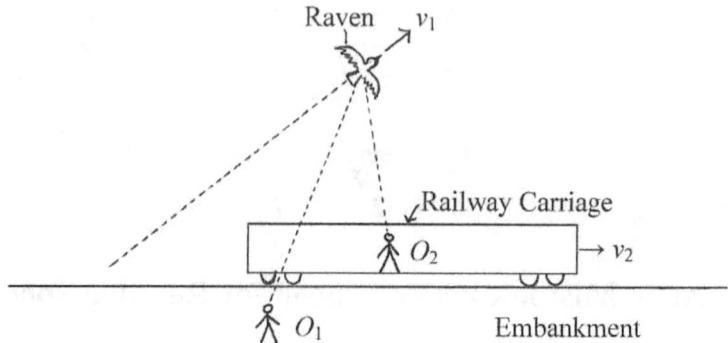

Fig. 7-1 v_1 is the proper velocity of the raven, and v_2 is the proper velocity of the railway carriage in the given environment.

The proper velocity of the raven with respect to the second inertial environment (the room inside the train, which is in motion with respect to the first inertial system, namely the embankment, at the uniform speed v_2) is constant (see underlined part #2 in the above citation) independently of the motion/position of observer O_2 or other observers (not shown) on the train.

Obviously, Einstein is talking about proper speed/velocity of the raven with respect to two different kinds of inertial environments—the embankment and the room of the train in motion.

The relative speed of the raven with respect a specific observer on the embankment is not constant. Normally (=unless observer O_1 is aligned with the line of motion of the raven), the speed of the raven relative to observer O_1 changes continuously as the raven moves. Likewise, the speed of the raven relative to observer O_2 on the train changes continuously unless the line of motion of the train coincides with the line of motion of the raven.

Normally, in reality, the raven and observer O_1 are in a 2-D situation; and the raven and observer O_2 are in a 3-D situation (see the definitions of 2-D situation and 3-D situation in **Chapter 3**).

If an observer sits in the line of motion of a moving object (this is a 1-D situation), the relative speed between the two is constant. We can see that Einstein considers relative speed *in terms of 1-D situation* regardless

Special Relativity

of whether the given situation was actually of 2-D or 3-D.

Modern physicists (relativists) use their own terms *velocity* and *proper velocity* whose meanings are totally different from mine. Wikipedia explains relativists' terms *velocity* and *proper velocity* as follows: [2]

> In relativity, proper velocity, also known as celerity, is an alternative to velocity for measuring motion. Whereas velocity relative to an observer is distance per unit time where both distance and time are measured by the observer, proper velocity relative to an observer divides observer-measured distance by the time elapsed on the clocks of the traveling object. Proper velocity equals velocity at low speed. Proper velocity at high speed, moreover, retains many of the properties that velocity loses in relativity compared with Newtonian theory.

The above statement sounds like the language of aliens for readers who are not familiar with relativity. I explain the above statement for readers as follows:

- The relativists' term *velocity* is similar to my term *proper velocity/speed*, which is observer-free.
- The relativists' term *proper velocity* is a kind of relative speed based on the assumption that distance contraction and time dilation really occur.
- Both relativists' terms *velocity* and *proper velocity* are one-dimensional (1-D) in concept; both terms are defined on the assumption that two involved objects are situated or moving on the same straight line with one party (observer) being stationary on the straight line.
- The act of explaining or supporting relativity using the relativists' terms of velocity and proper velocity, which are based on the assumption that distance contraction and time dilation really occur, is circular reasoning.

I do not know when and by whom the relativists' term **proper velocity** or **celerity** was coined. Einstein did not use the term **proper velocity** in his book *Relativity* (15th edition, 1952, Three River Press). ♦

8

Cosine Effect

In previous chapters we have already discussed the relative speed in 2-D situations. In this chapter, we will discuss the method of finding relative speed in term of *observation angle* (θ) in 2-D situations. An observation angle is the angle formed by two lines—the line of motion of a target object and the line that connects the target object and an observer (my term and definition). Let us see some examples:

In **Fig. 8-1**, object A moves along straight line EF at the uniform speed v in the direction of $E \rightarrow F$. E and F are points that are very far (∞) from each other. Observer O is off the line of motion A. Observation angle EAO (= θ) changes continuously as object A moves.

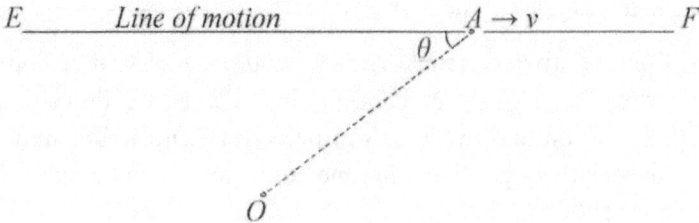

Fig. 8-1 Observation Angle EAO (= θ) changes continuously as object A moves.

We can find the relative speed of A (see **Fig. 8-1**) with respect to observer O in terms of observation angle (θ).

In **Fig. 8-2** (see next page), object A has moved along straight line EF from point A_1 to point A_2 at the speed v for Δt. If Δt is very small

Special Relativity

($\lim_{\Delta t \to 0}$), A_1A_2 ($= v\Delta t$) also becomes very small. Then, OA_1 and OA_2 become parallel with each other. OA_1 and OA_2 are shown as a curved dotted line in the diagram but they are actually straight parallel lines. Then observation angle EA_1O equals observation angle EA_2O. M is the point at which the perpendicular from A_1 to OA_2 meets OA_2.

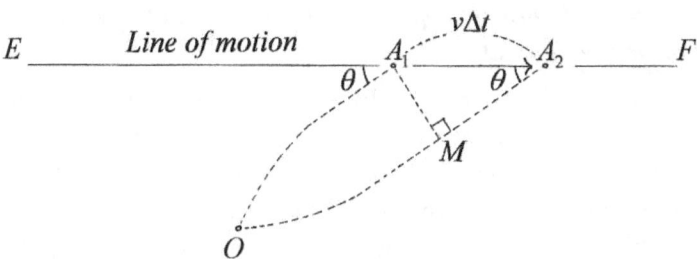

Fig.8-2 If Δt is very small, angle $EA_1O =$ angle EA_2O.

The *change* of "distance AO" during given time Δt is not A_1A_2 but $OA_2 - OA_1$. (Most physicists do not understand this!) Therefore, v_{AO} (= the speed of A relative to O) during given time Δt is $\dfrac{OA_2 - OA_1}{\Delta t}$.

Since $OA_2 = OM + MA_2$, and $OM = OA_1$,

$$v_{AO} = \frac{OA_2 - OA_1}{\Delta t} = \frac{OM + MA_2 - OM}{\Delta t} = \frac{MA_2}{\Delta t}.$$

(Since $MA_2 = A_1A_2 \cos\theta$ and $A_1A_2 = v\Delta t$,)

$$v_{AO} = \frac{MA_2}{\Delta t} = \frac{v\Delta t \cos\theta}{\Delta t} = v\cos\theta. \tag{4}$$

Eq. (4) is the speed equation of A relative to O at *observation angle* θ.

How do we know whether **Eq. (4)** is correct? We can prove **Eq. (4)** by replacing θ with several specific values as follows:

If $\theta = 0°$, $v_{AO} = v\cos\theta = v$.

(This is when object A is at point F which is very far from O.)

If $\theta = 180°$, $v_{AO} = v\cos\theta = -v$.

(This is when object A is at point E which is very far from O.)

If $\theta = 90°$, $v_{AO} = v\cos\theta = 0$.

(This is when A passes the point <not shown in Fig. 8-2> at which the perpendicular from O to EF meets EF. At this moment, the observation angle (θ) of A with respect to O is 90°.)

The above results show that **Eq. (4)** is consistent with **Eqs. (2)** and **(3)**. Note that **Eq. (4)** lacks r factor and t factor. **Eq. (4)** is an *observation–angle-priority equation* whereas **Eqs. (2)** and **(3)** are *time-factor-priority equations*.

I found **Eq. (4)** in 2004. [1] But until 2008 I did not know that **Eq. (4)** had already been known by the name of *cosine effect*. So the term *cosine effect* is not mine. In 2008, I found in the Internet that Donald Sawicki explains cosine effect in his manual for the US traffic officers. [2] Sawicki's method of finding the relative speed of vehicles with respect to observation angle is different from mine. Sawicki's term *angle* is *observation angle* in my terminology. Sawicki uses the term *actual speed* instead of *proper speed*. The term a*ctual speed* is better than *proper speed* because vehicles are usually not in uniform linear motion (see the definition of *actual speed* in **Chapter 1**).

When it comes to the cosine effect equation ($v\cos\theta$), Sawicki's and mine are the same. Sawicki explains cosine effect in his manual as follows: [2] The words in parenthesis are Sawicki's. The asterisk (*) is mine.

> Police microwave and laser radars measure the relative speed a vehicle is approaching (or receding) the radar. If a vehicle is traveling directly (collision course) at the radar, the relative speed is actual speed. If the vehicle is not traveling directly toward (or away) the radar but slightly off to avoid a collision, the relative speed with respect to the radar is slightly lower than actual speed. The phenomenon is called the Cosine Effect because the measured speed is directly related to the cosine of the angle (alpha) between the radar and vehicle direction of travel (see the figure* below). The greater the angle, the greater the speed error (the lower the measured speed). A cosine angle of 90° has 100% error (speed measures 0).

(*Readers can see the figure in the following Internet address: http://www.copradar.com/preview/chapt2/ch2d1.html.)

Special Relativity

Sawicki says, "A cosine angle of 90° has 100% error (speed measures 0)" because the relative speed of the vehicle with respect to the officer (who holds the speed detector) is zero (0) at the moment the observation angle is 90°. The relative speed of a vehicle appears in the speed detector is less than the actual (or proper) speed of the vehicle because the absolute value of $\cos\theta$ is always less than 1 except when $\theta = 0°$ or 180° ($\cos 0° = 1$, and $\cos 180° = -1$).

When it comes to the safety issue, observation angle 180° (this is when observer O is on straight line EF; see **Fig. 8-2**) should be avoided because this angle is a *collision course* in Sawicki's term. Observation angle 0° is alright.

The actual speed (or proper speed) of a vehicle appears on the speedometer of the vehicle. The speed appears on the speed gun (or speed detector) is the relative speed of the vehicle with respect to the speed gun. The closer the observation angle gets to 0° (or 180°), the closer the relative speed of the vehicle with respect to a speed detector becomes to the actual speed (or proper speed) of the vehicle in motion on the road. The speed appears on the speed detector (= $v\cos\theta$) is always smaller than the actual speed (v) of the vehicle as long as observation angle is 0° < θ < 180°. Therefore, if a traffic officer is to find as many speeders as possible, the officer (who hold the speed gun) needs to stand as closer to the road side (in order to minimize the observation angle) as possible. The father the speed detector gets from the road (or the line of motion of a car), the closer the observation angle gets to 90° and the smaller the speed of appears on the speed detector gets.

What do the cosine effect and speed detectors have to do with the theory of relativity? Cosine effect shows relativists confuse proper speed with relative speed. Speed detectors disprove relativity in the street and traffic officers confirm this every day. Only theoretical physicists in lofty ivory towers close their eyes and do not see what happens in the street. We do not need expensive particle accelerators to prove or disprove relativity.

Almost 100% of the physicists (both relativists and dissenters) I have contacted to date (from 2001 to 2014) disregard what speed guns and cosine effect show. Cosine effect, calculus, and speed the speed gun are the enemies of relativity. Relativists even do not know whether these are their enemies.

How Speed Gun Works
(Source: http://en.wikipedia.org/wiki/Radar_gun)

Speed guns use Doppler radar to perform speed measurements. Radar speed guns, like other types of radar, consist of a radio transmitter and receiver. They send out a radio signal in a narrow beam, then receive the same signal back after it bounces off the target object. Due to a phenomenon called the Doppler Effect, if the object is moving toward or away from the gun, the frequency of the reflected radio waves when they come back is different from the transmitted waves, and from that difference the radar speed gun can calculate the object's speed.

The target object's speed v is proportional to the difference in frequency Δf between the outgoing and the reflected radio waves:

$$\frac{v}{c} = \frac{\Delta f}{f}$$

where f is the frequency of the outgoing radio waves and c is the speed of light.

After the returning waves are received, a signal with a frequency equal to this difference is created by mixing the received radio signal with a little of the transmitted signal. Just as when two different musical notes are played together they create a "beat note" at the difference in frequency between them, when the two radio signals are mixed they create a "beat" signal (called a heterodyne) at the difference in frequency between the outgoing and reflected waves. The circuit then converts this frequency to a number by counting the number of cycles of the signal in a fixed time interval using a digital counter, and displays the number on a digital display as the object's speed.

It is important that the radio waves leave the gun in a narrow beam that doesn't spread out much, so that the gun will get a return only from the vehicle it is aimed at, with no chance of receiving a false return from nearby objects or vehicles. To create a narrow beam with an antenna small enough to fit in a handheld gun, radar speed guns use high frequency radio waves in the microwave range. X band (8 to 12 GHz) guns are becoming less common due to the fact the beam is strong and easily detectable. Also, most automatic doors utilize radio waves on X band can possibly affect the readings of police radar. As a result *K* band (18 to 27 GHz) and *Ka* band (27 to 40 GHz) are most commonly used by police agencies.

Cosine effect or **Eq. (4)** is completely compatible and consistent with **Eqs. (2)** and **(3)**. We can confirm this in the following exercise:

Exercise 8-1

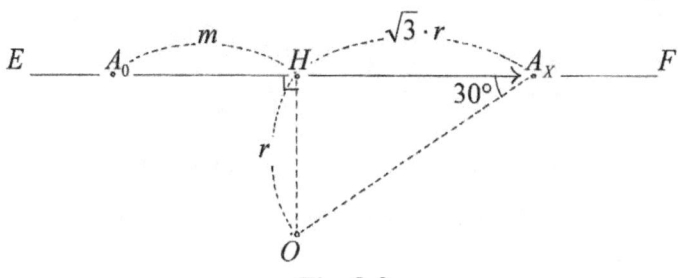

Fig. 8-3

In the diagram above, object *A* started from point A_0 and moves along straight line *EF* at the uniform speed *v* in the direction of *E*→*F*. A_0 is distance *m* from *H* toward *E*. *H* is the point at which the perpendicular from *O* to *EF* meets *EF*. Observer *O* is distance *r* from *EF*. That is, *OH* = *r*. Point A_X is distance $\sqrt{3}\,r$ from *H*. That is, angle $HA_XO = 30°$.

Question:

Find the speed of *A* relative to *O* at the moment *A* passes A_X by using **Eqs. (2), (3),** and **(4)** to see whether **Eqs. (2), (3),** and **(4)** are consistent and compatible with each other.

Solution:

Since *A* starts from A_0, the time *A* passes A_X will be $\dfrac{(m+\sqrt{3}\cdot r)}{v}$.

At the moment $t = \dfrac{(m+\sqrt{3}\cdot r)}{v}$, the relative speed of *A* with respect to *O*, according to **Eq. (2)**, is

$$Y' = \frac{v(vt-m)}{\sqrt{(vt-m)^2 + r^2}} = \frac{v(m+\sqrt{3}\cdot r - m)}{\sqrt{(m+\sqrt{3}\cdot r - m)^2 + r^2}} = \frac{\sqrt{3}}{2}r.$$

If A starts from point H, the time A passes A_X will be $t = \frac{\sqrt{3}\cdot r}{v}$. Then, according to **Eq. (3)**, the relative speed of A with respect to O at the moment $t = \frac{\sqrt{3}\cdot r}{v}$ is

$$Y' = \frac{v(vt)}{\sqrt{(vt)^2 + r^2}} = \frac{v\sqrt{3}\cdot r}{\sqrt{(\sqrt{3}\cdot r)^2 + r^2}} = \frac{\sqrt{3}}{2}v.$$

The observation angle of A with respect to O at the moment A passes A_X is 30°. According to cosine effect or **Eq. (4)**, the speed of A relative to O ($=v_{AO}$) is

$$v_{AO} = v\cos 30° = \frac{\sqrt{3}}{2}v.$$

The above results show that **Eqs. (2), (3),** and **(4)** are consistent and compatible with each other. We can confirm that **Eqs. (2), (3),** and **(4)** are consistent with each other by applying specific angles such as 0°, 45°, 60°, 90°, 180°, etc. to the observation angle θ.

If we use the cosine table, we can find the relative speed of A with respect to O at any value of θ. For example, if $\theta = 103.3°$, $v_{AO} = v\cos 103.3° = -0.2300v$. Note that $\cos 103.3° = \cos(180°-103.3°) = -\cos 76.7° = -0.2300$. The negative sign means that distance AO decreases (−) at the given angle.

Exercise 8-2

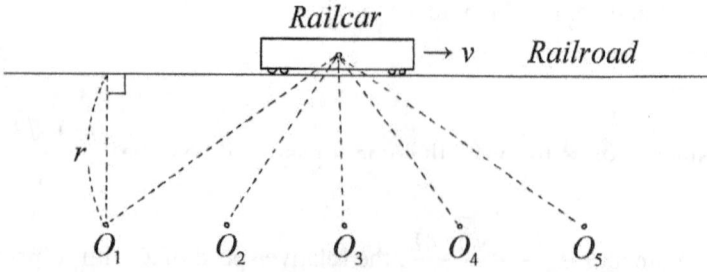

Fig. 8-4 Each observer is distance r from the railroad.

In **Fig. 8-4**, a railcar moves at the uniform speed v on a straight railroad, and five observers O_1, O_2, O_3, O_4, and O_5 are deployed along the embankment with the distance of each observer from the railroad being r.

Question:

Prove or disprove whether the relative speed of the railcar is the same (v) and uniform for all the observers at any moment.

Relativistic solution:

The relative speed of the railcar is the same (v) and constant for all the observers at any moment.

Ahn's solution

The observation angles of the railcar with respect to the observers are all different from each other at any moment. Therefore, according to **Eq. (4)** ($v_{AO} = v \cos \theta$), the *relative speeds* of the railcar with respect to the observers are all different from each other at any moment. Yet the *proper speed* of the railcar is the same and uniform irrespectively of the position or state of motion of the observers.

In this exercise (**Fig. 8-4**), the railcar is treated as a point for the convenience's sake. But in fact a train or a railcar is not a point. It has a considerable length. Therefore, the speed of the head portion of the railcar relative to one of the observers on the embankment is different from the speed of the tail part of the railcar relative to the same observer at any moment (due to the different observation angles). I explained this using calculus to the members of the NPA (Natural Philosophy Alliance) at 2005 NPA Conference, and no one understood. Some board members thought I was beside myself; one of them said to me that the train would be broken in two parts if my argument were true. Most of the NPA members were doctors or masters degrees holders.

Exercise 8-3

In **Fig. 8-5** (see next page), car A moves along straight road EF at the uniform speed v in the direction of $E \rightarrow F$. Three observers O_1, O_2, and O_3 are so deployed that the perpendiculars from these three observers to EF meet EF at the same point H. That is, angle $EHO_1 =$

angle EHO_2 = angle EHO_3 = 90°.

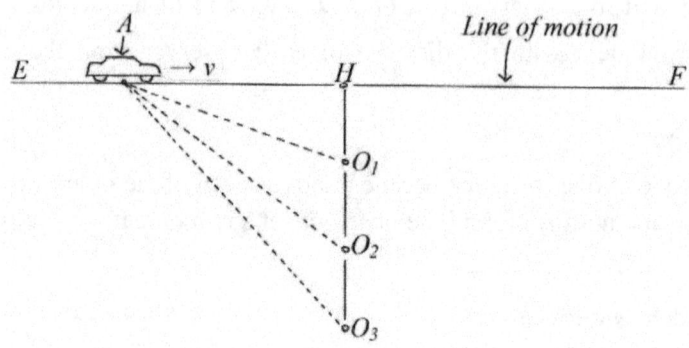

Fig. 8-5 Angle EHO_1=angle EHO_2=angle EHO_3= 90°.

Question:

Prove or disprove that the speed of the car is the same (v) relative to the three observers.

Relativistic solution

The relative speed of the car is v and constant for the three observers.

Ahn's Solution

Observation angles EAO_1, EAO_2, and EAO_3 are different from each other at any moment except the moment the car passes point H, at which the observation angle is 90° for the three observers. When the car passes point H, the speeds of the car relative to the three observers are all zero (0).

Cosine Effect and Speed Detector

In 2001, I illustrated, using **Eq. (3)**, that the relative speed of an object with respect to an observer, who is off the line of motion of the object, changes continuously as the object moves. [3] In 2004, I found **Eq. (4)** ($V_{AO} = v\cos\theta$) [1]. In 2006, I proved mathematically that **Eq. (4)** is compatible with **Eq. (2)** and **Eq. (3)** as shown in **Exercise 8-1**. [4] But until 2008 I had not confirmed whether radar guns work as I explained with **Eq. (4)**.

On September 17, 2008, I came to the Hamden Police Office

Special Relativity

(Hamden, Connecticut), which was only two blocks away from my house, with a simplified illustration (drawing) of my argument explaining the relationship between observation angles and speed detectors. I was told by an officer that my argument was correct. The officer added that more accurate *laser* was used in modern days instead of the old type radar.

The next day, Sept. 18, 2008, I came to the New Haven Police Department (New Haven, Connecticut) to re-confirm my argument. But this time I was told by an officer that my thought was not correct. The officer said that the speed of a car is the same for any speed detectors regardless of position the speed gun. I thought that not all officers knew the principle of speed gun. In the evening that day (Sept, 18, 2008) I found on the Internet (www.copradar.com) that speed detectors work according to **Eq. (4)** and that this equation was already known by the name of *cosine effect*. That means I am not the one who first found the *cosine effect*. On the same Internet (www.copradar.com), Donald Sawicki explains the *cosine effect* in detail in his manual titled *Police Traffic Radar Handbook* (1999).

Observation Angle and Line of Motion

Observation angle or angle of observation (θ) is an angle formed by two lines—the line of motion of an object and the line that connects the object with an observer. The line of motion is a straight line; see line *EF* in the diagram below.

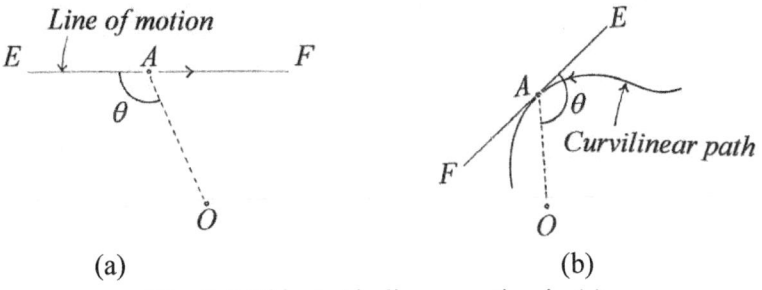

(a) (b)

Fig. 8-6 Object *A* in linear motion in (a).
Object *A* is in curvilinear motion in (b)

In diagram (a) (see **Fig. 8-6**), object *A* is in uniform motion on a straight line *EF* in the direction $E \rightarrow F$. Observer *O* sits off the straight line *EF*. *EF* is the line of motion of *A*, and angle *EAO* (not angle *FAO*) is the observation angle (θ) of object *A* with respect to observer *O*. Note that the direction of motion of object *A* is $E \rightarrow F$, not $F \rightarrow E$. Since object *A* moves, observation angle *EAO* changes accordingly. The relative speed between *A* and *O* is the proper speed of *A* times cosine of angle *EAO* ($=v\cos\theta$).

In graph (b), object *A* is in curvilinear motion along a curvilinear path. Straight line *EF* is the tangent at the present position of *A* on its curvilinear path. Angle *EAO* (θ) is the observation angle of *A* with respect to observer *O* at the given moment. The relative speed between *A* and *O* is the tangent speed of *A* at the given moment times the cosine of angle *EAO*. ◆

Special Relativity

9

The Case in Which Two Lines of Motion Meet

In **Fig. 9-1**, two objects A and B started from common origin O at the same time at the uniform speeds v_A and v_B respectively maintaining intersection angle $AOB = \theta$. Origin O is stationary in the given inertial frame of reference. Let us find v_{AO}, the relative speed of A with respect to B *or vice versa*, from the relativistic point of view and then from the Galilean point of view.

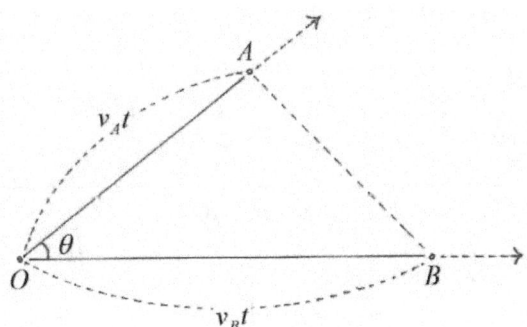

Fig. 9-1 Angle $AOB = \theta$.

Relativistic solution:
Relativity deals with 1-D situations, neither 2-D nor 3-D situations. Since this is a case of 2-D situation, relativists cannot solve this problem.

Galilean solution:
If we are to find the relative speed between A and B, we need to find the equation that represents distance AB, and then we find the derivative

101

of the distance equation. So let us find the equation that represents distance AB.

In **Fig. 9-2**, H is the point at which the perpendicular from A to OB meets OB. AHB = 90°.

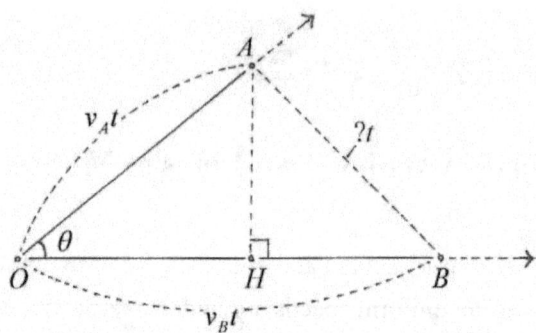

Fig. 9-2 Angle $AHB = 90°$.

Let distance AB be Y. (Since AHB is a right triangle,)

$$AB = Y = \sqrt{(AH)^2 + (HB)^2}$$

(Since $AH = v_A t \sin\theta$, and $HB = OB - OH = v_B t - v_A t \cos\theta$,)

$$= \sqrt{(v_A t \sin\theta)^2 + (v_B t - v_A t \cos\theta)^2}$$

$$= t\sqrt{(v_B)^2 + (v_A)^2(\sin^2\theta + \cos^2\theta) - 2v_A v_B \cos\theta}$$

(Since $\sin^2\theta + \cos^2\theta = 1$,)

$$= t\sqrt{(v_A)^2 + (v_B)^2 - 2v_A v_B \cos\theta}. \qquad (5)$$

Eq. (5) is the equation that represents distance AB with respect to time t. The derivative of **Eq. (5)** with respect to time t is

$$Y' = \frac{dy}{dt} = \sqrt{(v_A)^2 + (v_B)^2 - 2v_A v_B \cos\theta}. \qquad (6)$$

Eq. (6) represents the relative speed of A with respect to B. Note that there is no time factor t in **Eq. (6)**. This means that the relative speed of A with respect to B is constant (as long as θ is fixed).

Special Relativity

Fig. 9-1 (and **Fig. 9-2**) is the case in which angle *ABO* is less than 90°. We find the same results when angle *ABO* is equal to or less than 90°.

How do we know whether **Eq. (6)** is correct? We can prove **Eq. (6)** by replacing v_A, v_B, and θ with some actual values as follows:

If $\theta = 0°$, $v_A \neq 0$, and $v_B \neq 0$,

$$Y' = \sqrt{(v_A)^2 + (v_B)^2 - 2v_A v_B \cos 0°}$$

(Since $\cos 0° = 1$,)

$$= \sqrt{(v_A)^2 + (v_B)^2 - 2v_A v_B}$$

$$= |v_A - v_B| \text{ or } v_A \sim v_B.$$

(This is when *A* and *B* move on the same straight line in the same direction. This is a 1-D situation.)

If $\theta = 90°$, $v_A \neq 0$, and $v_B \neq 0$,

$$Y' = \sqrt{(v_A)^2 + (v_B)^2 - 2v_A v_B \cos 90°}$$

(Since $\cos 90° = 0$,)

$$= \sqrt{(v_A)^2 + (v_B)^2}.$$

If $\theta = 180°$, $v_A \neq 0$, and $v_B \neq 0$,

$$Y' = \sqrt{(v_A)^2 + (v_B)^2 - 2v_A v_B \cos 180°}$$

(Since $\cos 180° = -1$,)

$$= \sqrt{(v_A)^2 + (v_B)^2 + 2v_B v_B} = v_A + v_B.$$

(This is when two objects *A* and *B* move in the opposite directions of each other on the same straight line.)

If $v_A = 0$ and $v_B \neq 0$,

$$Y' = \sqrt{(v_B)^2} = v_B.$$

(This is 1-D situation; *A* stays at point *O*.)

If $v_A \neq 0$ and $v_B = 0$,

$$Y' = \sqrt{(v_A)^2} = v_A.$$

(This is 1-D situation; B stays at point O.)

Therefore, we can see that **Eq. (6)** is correct.

Galilean solution is versatile and makes sense. That is, **Eq. (6)** is good for any value of v_A, v_B, and θ.

Fig. 9-1 is the case in which two objects (A and B) start from a common starting point (O) at the same time. But two objects need not start from a common point at the same time. The case in which two objects start from different origins respectively and pass an intersection point (O) at different times is discussed in **Exerces-A1-4** in **Appendix-1**.

♦

Special Relativity

10

The Case of Parallel Motion

If two objects A and B move on the same straight line at the uniform speeds v_A and v_B respectively, the relative speed between the two is simply $|v_A - v_B|$ or $v_A \sim v_B$. This is a 1-D situation.

What if two objects A and B move on the different straight lines that are parallel to each other? This is a 2-D situation. In the diagram below, object A moves at uniform speed v_A on a straight line $E_A F_A$ in the direction $E_A \rightarrow F_A$. Object B moves at a uniform speed v_B on a straight line $E_B F_B$ in the direction $E_B \rightarrow F_B$. Two straight lines $E_A F_A$ and $E_B F_B$ are parallel to each other.

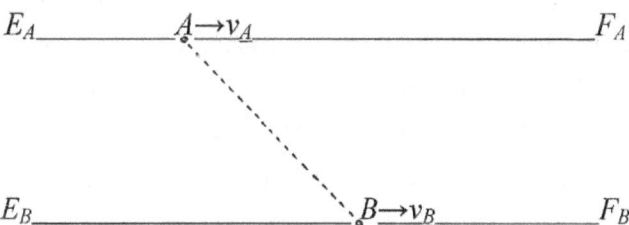

Fig. 10-1 $E_A F_A$ and $E_B F_B$ are parallel to each other.

In the above case, most (probably all) physicists (both relativists and dissidents) have thought that the relative speed between A and B is $v_A \sim v_B$. They are wrong. Note that A and B are in a 2-D situation, not 1-D situation. Let us find the relative speed between two objects A and B in the following exercise.

105

Exercise 10-1

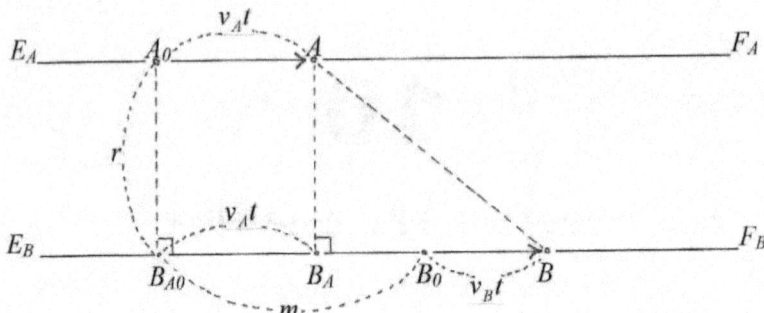

Fig. 10-2 $E_A F_A$ and $E_B F_B$ are parallel to each other.

In the diagram above, two straight lines $E_A F_A$ and $E_B F_B$ are parallel to each other. The distance between the two parallel lines is r. Objects A started from point A_0 at time $t = 0$ and moves along $E_A F_A$ in the direction $A_0 \rightarrow F_A$ at the uniform speed v_A. That is, $A_0 A = v_A t$. Object B started from point B_0 at time $t = 0$ and moves along $E_B F_B$ in the direction $B_0 \rightarrow F_B$ at the speed v_B. $B_0 B = v_B t$. B_{A0} is the point at which the perpendicular from A_0 to $E_B F_B$ meets $E_B F_B$. $A_0 B_{A0} = r$. Since A_0 is stationary, B_{A0} is also stationary. B_A is the point at which the perpendicular from A to $E_B F_B$ meets $E_B F_B$. $AB_A = r$. Since A moves at the speed v_A, B_A also moves at the same speed v_A along $E_B F_B$ in the direction $B_A \rightarrow F_B$. $B_{A0} B_0 = m$.

Find the following:

(1) The equation that represents distance AB and the equation that represents the relative speed of A with respect to B or vice versa.

(2) The relative speed of A with respect to B at the moment AB is perpendicular to both $E_A F_A$ and $E_B F_B$. (Assume that $v_A > v_B$.)

Solution:

Special Relativity

(1) Let distance AB be Y. Since $A B_A B$ is a right triangle,

$$AB = Y = \sqrt{(B_A B)^2 + (AB_A)^2}$$

(since $B_A B = |m + v_B t - v_A t|$, and $AB_A = r$,)

$$= \sqrt{(m + v_B t - v_A t)^2 + r^2}. \tag{7}$$

Eq. (7) is the equation that represents distance AB.

The derivative of Y with respect to time t is

$$Y' = \frac{(v_A - v_B)[(v_A - v_B)t - m]}{\sqrt{[(v_A - v_B)t - m]^2 + r^2}}. \tag{8}$$

Eq. (8) represents the relative speed of A with respect to B or vice versa.

How do we know whether **Eq. (8)** is correct?

We can prove **Eq. (8)** by checking the following cases.

If $v_A = v_B$, $Y' = \dfrac{(v_A - v_B)[(v_A - v_B)t - m]}{\sqrt{[(v_A - v_B)t - m]^2 + r^2}} = 0$.

(In this case, distance AB does not change.)

If $v_A \neq 0$, $v_B \neq 0$, and $r = 0$,

$$Y' = \frac{(v_A - v_B)[(v_A - v_B)t - m]}{\sqrt{[(v_A - v_B)t - m]^2 + r^2}} = |v_A - v_B|.$$

(This is a 1-D situation; A and B move on the same straight line in the same direction.)

If $v_A \neq 0$, $v_B = 0$, and $r = 0$, (This is a 1-D situation.)

$$Y' = \frac{(v_A - v_B)[(v_A - v_B)t - m]}{\sqrt{[(v_A - v_B)t - m]^2 + r^2}} = \frac{(v_A)(v_A t - m)}{\sqrt{(v_A t - m)^2}} = v_A.$$

If $v_A = 0$, $v_B \neq 0$, and $r = 0$, (This is a 1-D situation.)

$$Y' = \frac{(v_A - v_B)[(v_A - v_B)t - m]}{\sqrt{[(v_A - v_B)t - m]^2 + r^2}} = v_B.$$

If $v_A \neq 0$, $v_B = 0$, and $r \neq 0$,

$$Y' = \frac{(v_A - v_B)[(v_A - v_B)t - m]}{\sqrt{[(v_A - v_B)t - m]^2 + r^2}} = \frac{v_A(v_A t - m)}{\sqrt{(v_A t - m)^2 + r^2}}.$$

(This is the same form as **Eq. (2)** in **Chapter 4**.)

If $v_A \neq 0$, $v_B = 0$, $r \neq 0$, and $m = 0$,

$$Y' = \frac{(v_A - v_B)[(v_A - v_B)t - m]}{\sqrt{[(v_A - v_B)t - m]^2 + r^2}} = \frac{(v_A)^2 t}{\sqrt{(v_A)^2 + r^2}}.$$

(This is the same form as **Eq. (3)** in **Chapter 4**.)

If $v_A = 0$, $v_B \neq 0$, and $r \neq 0$,

$$Y' = \frac{(v_A - v_B)[(v_A - v_B)t - m]}{\sqrt{[(v_A - v_B)t - m]^2 + r^2}} = \frac{v_B(v_B t + m)}{\sqrt{(v_B t + m)^2 + r^2}}.$$

(This equation is *not* the same as **Eq. (2)**. In this case, AB will never be perpendicular to the line of motion of B.)

Therefore, we can see that **Eq. (8)** is correct.

(2) From $v_A t = m + v_B t$, the time AB becomes perpendicular to both parallel lines is $t = \dfrac{m}{v_A - v_B}$.

If $t = \dfrac{m}{v_A - v_B}$ in **Eq. (8)**,

$$Y' = \frac{(v_A - v_B)[(v_A - v_B)t - m]}{\sqrt{[(v_A - v_B)t - m]^2 + r^2}} = \frac{(v_A - v_B)(m - m)}{\sqrt{(m - m)^2 + r^2}} = \frac{0}{r} = 0.$$

(This is when the observation angle of A with respect to B becomes 90°. Note that if $v_A \leq v_B$, AB will never be perpendicular to both parallel lines.)

Fig. 10-2 is the case in which two objects A and B move in the *same direction* on parallel lines. The case in which two objects move in the *opposite direction* on parallel lines is illustrated in **Exercise A1-1** in **Appendix-1**. ♦

Special Relativity

11

The Case of Accelerated Motion

We have learned in **Chapters 4, 8**, and **10** that if observation angle becomes 90°, the relative speed between two involved objects becomes zero ($v = 0$). This is the case one or both objects are in uniform motion in 2-D situation. What if any one or both objects are in non-uniform motion in 2-D situation? Will the phenomenon of $v = 0$ occur at the moment observation angle becomes 90°? Let us see.

Exercise 11-1

In **Fig. 11-1**, object A started at A_0 at time (t) = 0 and moves along straight line EF in the direction $E \rightarrow F$ at the speed "at" where a is acceleration and t is time. Then, $A_0 A = at^2/2$ A_0 is distance m from H. Stationary observer O is distance r from EF. H is the point at which the perpendicular from O to EF meets EF.

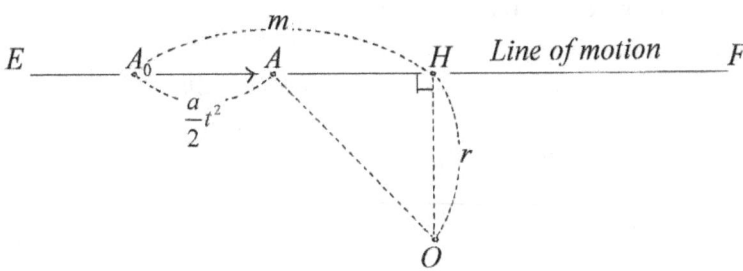

Fig. 11-1 Distance $A_0 A = at^2 / 2$.

Questions:

(1) Find the equation that represents the distance AO.

(2) Find the equation that represent the relative speed between A and O.

(3) Find the relative speed of A with respect to O at the moment A passes point H.

Solution:

(1) Let distance AO be Y. Since AOH is a right triangle,

$$Y = AO = \sqrt{(AH)^2 + (OH)^2}$$

$$= \sqrt{(A_0H - A_0A)^2 + (OH)^2}$$

(Since $|A_0H - A_0A| = \left|m - \frac{1}{2}at^2\right|$, and $OH = r$,)

$$= \sqrt{\left(\frac{1}{2}at^2 - m\right)^2 + r^2} . \quad (9)$$

Eq. (9) represents distance AO with respect to time t.

(2) The derivative of **Eq. (9)** with respect to time t is

$$Y' = \frac{dy}{dt} = \frac{a^2t^3 - 2amt}{2\sqrt{\frac{1}{4}a^2t^4 - amt^2 + m^2 + r^2}} . \quad (10)$$

Eq. (10) is the equation that represents the relative speed of A with respect to O.

(3) We can find the speed of A with respect to O at the moment A passes point H by using **Eq. (10)** as follows:

From $\frac{1}{2}at^2 = m$, the time A passes point H is $t = \sqrt{2m/a}$.

If $t = \sqrt{2m/a}$ in **Eq. (10)**,

$$Y' = \frac{a^2t^3 - 2amt}{2\sqrt{\frac{1}{4}a^2t^4 - amt^2 + m^2 + r^2}} = 0.$$

[Note that numerator $a^2t^3 - 2amt = at(at^2 - 2m) = at(2m - 2m) = 0$.]

However, the *actual speed* (= instantaneous rectilinear speed) of A in its line of motion at the moment A passes H is not zero (0) but $at =$

Special Relativity

$a \times \sqrt{2m/a} = \sqrt{2am}$.

Therefore, we can see that the third law of relative speed (see the box below) is true even when an object is in accelerated motion. The above exercise (**Exercise 11-1**) is the case in which an object is in non-uniform motion and an observer is stationary in the given inertial system. Observation angle 90° is formed only in 2-D situations. Therefore the third law of relative speed is applied exclusively to 2-D situations.

In a 3-D situation, two involved objects can (very rarely) be in the moment at which the observation angle becomes 90°. At this moment, the two involved objects are in 2-D situation instantaneously. At this moment, the relative speed between the two is zero (0) (see **Exercise 5** in **Appendix-1**).

In general, the third law of relative speed (see **Chapter 4**) holds regardless of the type of motion of one or both involved objects. We will check this in some more examples of non-uniform motion in the following chapters.

The Third Law of Relative Speed

At the moment the *observation angle* is 90°, the relative speed between two involved objects is zero (0) regardless of the types of motion or magnitudes of the speed of the two objects (see **Chapter 4**). ♦

12

The Case of Circular Motion

In **Fig. 12-1** below, artificial satellite A makes a round orbit around the earth with fixed distance r from the center (O) of the Earth. Straight line EF is the tangent at the present position of satellite A. The angular velocity of the satellite is ω ($=\dfrac{d\phi}{dt}$) where ϕ is radian and ω is the rate of change in the radian. The tangent speed (v) of the satellite on its obit is $r\dfrac{d\phi}{dt}$. In this case, what is the relative speed of the satellite with respect to the center (O) of the earth?

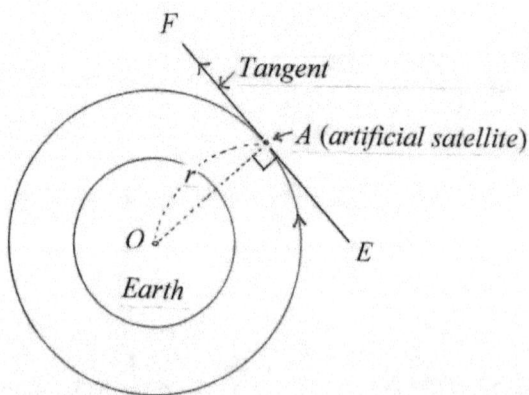

Fig. 12-1 Artificial satellite A revolves around the earth.

In **Fig. 12-1**, r or OA (the distance of A from O) does not change. Since there is no change in the distance between the satellite and the

Special Relativity

observer *O*, the relative linear speed between the two is *always* zero (0). Note that observation angle *EAO* is always 90°.

Most physicists think that the tangent speed v $(= r\dfrac{d\phi}{dt})$ of the satellite is the relative speed of the satellite with respect to an (or any) observer on the earth. They (physicists) are wrong! In this case, the angular speed or tangent speed of the satellite is a kind of proper speed which is independent of observers.

What relativity should deal with is the rate of change in the distance (= relative speed) between two objects. The change in the distance between *A* and *O* is always zero (0) regardless of the rotation speed (angular speed) or the tangent speed of the satellite (if we assume that artificial satellite *A* maintains a round orbit around).

What if observer *O'* (See **Fig. 12-2**) sits on the surface of the earth (instead of at the center of the earth)? (Satellite *A* is rotating in clockwise direction.) Will the relative speed between the satellite and observer *O'* also be zero (0) as in the case of **Fig. 12-1**?

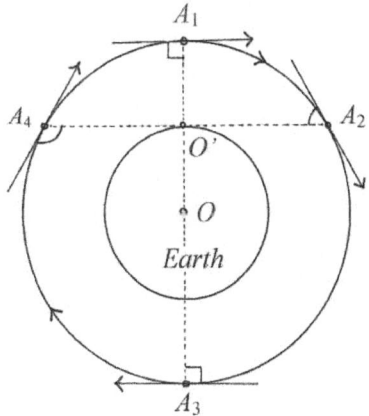

Fig. 12-2 Observer *O'* sits on the surface of the earth.

In **Fig. 12-2**, the observation angle (θ) of the satellite with respect to observer *O'* is not always 90° but changes periodically in the range of 0° $< \theta <$ 180°.

In the diagram, the observation angle becomes 90° at the moment satellite is at position A_1 and A_3. At each of these two moments, the relative speed between the satellite and the observer becomes zero (0).

The observation angle becomes a minimum when the satellite is at position A_2, and the maximum when at position A_4. A_2 and A_4 are the positions of the satellite when the observer sees the satellite on the horizon.

In reality, the orbits of satellites are elliptical and not round. In any event, the relative speed between a satellite and on observer on the surface of the earth is not uniform but changes continuously and periodically.

Relativists think that the orbital or tangential speed of a GPS satellite is the relative speed of the satellite with respect to an (any) observer on the earth. Orbital speed or tangent speed of satellite is a kind of proper speed which is independent of observers. This means that relativists do not know what relative speed is. Relativists are applying wrong speed to GPS (see **Chapter 14**). ♦

Special Relativity

13

The Case of Rotation Motion

In **Fig. 13-1**, a disc rotates at a uniform rate in its own plane about its center in *Galilean space K*. (The term *Galilean space* means free space. Assume that free space is absolute space.) The letter *K* is not shown in the diagram). Observer *O* is at the center of the rotating disc.

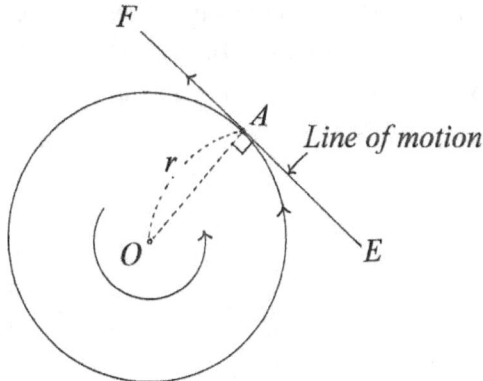

Fig. 13-1 Observer *O* is at the center of a rotating disc.

Object *A* is held fast on the edge of the rotating disc so that the object would not fall or slip from the edge of the disc. The radius of the disc is *r*. The angular speed of the disc is ω (= $\frac{d\phi}{dt}$ where ϕ is change in the angle and *t* is time), and the tangent speed of object *A* is *v* (= $r\frac{d\phi}{dt}$). Straight line *EF* is the tangent at the present position of *A*. *EA* is the line

115

of motion of A at the present position of A. Assume that the mass of object A does not affect the uniform rotation motion of the disc.

In the above situation, what is the relative speed of A with respect to O? In this case, the distance (r) between A and observer O is *always the same and* does not change regardless of the magnitude of the angular speed (ω) or the tangential speed of A. Therefore, the relative speed between A and O is always zero (0). Note that the observation angle EAO is always 90°.

In the above situation, not only distance OA but also the distance between any two fixed points on the disc does not change regardless of the rotation motion of the disc (because the disc is a solid body). Therefore the relative speed between any two fixed points or parts on the disc is always zero (0). If any readers have difficulty in understanding this, let us think about the case of the earth. The earth rotates at a considerable speed. Yet the distance between any two fixed points (or objects) on or inside the earth remains the same regardless rotation motion of the earth. (Let us assume here that the earth is a solid body for our thought experiment.) Then the relative speed between any two points on or inside the earth is always zero (0).

What if observer O' rests in the Galilean space K outside the rotating disc as shown in **Fig. 13-2**?

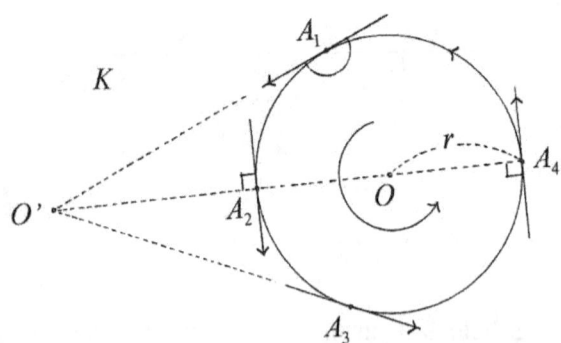

Fig. 13-2 Observer O' rests outside the disc in Galilean space K.

How the relative speed of A on the edge of the disc with respect to observer O'? In this case (see **Fig. 13-2**), the *observation angle* of object A with respect to observer O' changes periodically within the range of 0°

$\leq \theta \leq 180°$. That is, the relative speed between object A and O' changes continuously.

Four points A_1, A_2, A_3, and A_4 in **Fig. 13-2** are special positions of A in relation with observer O':

A_1 is the position of A at which the observation angle of A with respect to O' is 180°. At this moment, the relative speed of A with respect to O' is $v\cos 180° = -v$ or $-r\dfrac{d\phi}{dt}$. At this moment, distance $O'A$ *decreases* (−) at the rate of $r\dfrac{d\phi}{dt}$, the tangential speed of A on the rem of the disc. Hence the negative sign (−).

A_3 is the position of A at which the observation angle of A with respect to O' is 0°. At this moment, the relative speed of A with respect to O' is $v\cos 0° = v$ or $r\dfrac{d\phi}{dt}$.

A_2 and A_4 are the positions of A at each of which the observation angle of A with respect to O' is 90°. At each of the two positions, the relative speed between O' and A is $v\cos 90° = $ zero (0).

Summing up, the relative speed of A with respect O' changes continuously and periodically with the range of $0 \leq v_{AO} \leq r\dfrac{d\phi}{dt}$ where (v_{AO} is the relative speed between A and O and $r\dfrac{d\phi}{dt}$ is the tangential speed A on the rim of the disc).

Einstein did the same disc thought experiment as shown in **Fig. 13-2**. However, he explained that the speed of the object at the edge of a rotating disc is uniform when observed from the Galilean space K (see **Chapter 40**). [1] This means that what Einstein is talking about is the tangential speed ($r\dfrac{d\phi}{dt}$) of the object. This speed is a kind of proper speed which is independent of observers. Here again we find that what Einstein dealt with was not relative speed but proper speed.

Einstein's disc thought experiment is very important for relativists in the point that it is not only the foundation of general relativity but also the bridge that connects special relativity with general relativity. Since Einstein disc thought experiment has fundamental fallacy, both special and general theories of relativity are wrong (see **Chapter 40**). [1] ♦

Special Relativity

14

GPS and Relativity

The global positioning system (GPS) consists of 24 satellites whose semi-round orbital radius is about 26,600 km (16,500 mi). Each of these satellites makes a complete revolution around the earth every 12 hours. [1] (The average radius of the earth is about 6,370 km.) Relativists say that GPS constantly corrects information based on the speed of the satellites in order to reflect the relativistic effects (time dilation). The problem is that the speeds of the satellites relativists deal with for the information correction are not the *relative speeds* between the satellites and observers (receivers) on the earth.

Relativists say that the average orbital speed of the satellite is about 3.87km/s. [2] 3.87km/s is the orbital speed of the satellites obtained by assuming that the earth is an inertial frame of reference. The orbital speeds of GPS satellites are the proper speeds that are independent of observers on the earth or anywhere else. That is, the orbital speeds of the satellites are not the relative speed of the satellites with respect to the observers on the surface of the earth. The earth is not an inertial system. The earth is in very complicated motion.

If we assume that the earth is an inertial system and we also assume that the orbit of a GPS satellite is round, the relative speed between the satellites and the *center* of the earth is always zero (0) regardless of the ***orbital speed*** of the satellites or the rotational speed of the earth. This is because the distance between the satellite and the center of the earth remains the same. We have seen this in the previous chapters (**Chapters 12** and **13**). And the lineal distance between the satellite and an observer on the surface of the earth changes periodically. That is, the relative speed between the GPS satellite and the observer on the surface of the earth changes periodically from zero (0) to a certain value that is less than the orbital or tangential speed of the satellite in its orbit (confer **Chapters 12** and **13**).

The fact that relativistic correction of the GPS information is done based the orbital speeds (proper speeds) of the satellites and not the relative speed between the satellites and the observers on earth shows that relativistic correction of the GPS information is wrong.

Relativists say that GPS information is also corrected based on general relativity. But general relativity is based on the *equivalence principle* which is but the *misidentification* of gravity with inertial force. The equivalence principle is wrong: Gravity is one thing and inertial effect/force (from acceleration) is totally another (see **Chapter 36**).

No doubt relativists apply both special and general relativity to GPS the wrong way. Then why the *relativistic correction* of the GPS information seems to work all right? Relativists fumble with GPS information. There are so many factors to be considered in the GPS science: The motion/speed of the earth and that of the GPS satellites are affected by so many factors such the complicated motion of the rotation and revolution of the earth in space, the complicated variations of the earth's gravity due to the motion of the moon, the wobbling motion of the earth due to the motion and gravity of the moon, etc.

My position is that the speed (proper speed) of light is uniform with respect to absolute space and not to the earth or the observers on the earth. The earth itself is not an inertial frame of reference. The fact that the earth has a turbulent atmosphere makes the speed of light even more complicated one for both relativists and dissenters. Then the notion of the relativistic correction of GPS information sounds even more untenable.

Recently (on July 3, 2013, to be exact), I read a nice paper entitled *Does the GPS System Rely upon Einstein's Relativity?* (2012) by **Barry Springer**. The author earned B.S. and M.S. degrees in Electrical Engineering, and served in the U.S. Air Force in the Space Operations career field, directing satellite control systems and on-orbit spacecraft operations. He served as Commander of the GPS Master Control Station. A part of his paper reads as follows: [3] (The reference number [2] in the citation is Springer's. Asterisk is mine.)

There are mathematical purists who will insist that relativistic effects can be found in the GPS system that could be eliminated through corrections based on Einstein's relativity. But even they admit that "introducing the gamma factor makes a change of only 2 or 3 millimeters to the classical result." [2]* Such small error corrections, if even valid, would represent only a 0.1% adjustment in a typical navigation solution accurate to 2 or 3 meters.

We have found no need to be concerned about adjustments for Einstein's special or general relativity effects with regard to routine GPS operations – they are simply not required.

----

*[2] Tom Van Flandern, "Open Questions in Relativistic Physics, Aperion," 81-90, 1998.

Springer's whole paper can be found at the following address:

http://www.worldnpa.org/site/member/?memberid=2323&subpage=abstracts. ♦

15

Relativity of simultaneity

Einstein explains ***the relativity of simultaneity*** in his book *Relativity* (fifteenth ed.) as follows: [1] (The figure number is reset for this book.)

> We suppose a very long train traveling along the rails with the constant velocity v and in the direction indicated in **Fig. 15-1**. People traveling in this train will with advantage use the train as a rigid reference-body (co-ordinate system); they regard all events in reference to the train.

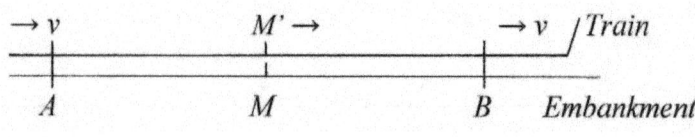

Fig. 15-1

> Then every event which takes place along the line also takes place at a particular point of the train. Also the definition of simultaneity can be given relative to the train in exactly the same way as with respect to the embankment. As a natural consequence, however, the following question arises:
>
> Are two events (*e.g.* the two strokes of lightning A and B) which are simultaneous *with reference to the railway embankment* also simultaneous *relatively to the train*? We shall show directly that the answer must be in the negative.
>
> When we say that the lightning strokes A and B are simultaneous with respect to the embankment, we mean: the rays

Special Relativity

of light emitted at the places A and B, where the lightening occurs, meet each other at the mid-point M of the length $A{\rightarrow}B$ of the embankment. But the events A and B also correspond to positions A and B on the train. Let M' be the mid-point of the distance $A{\rightarrow}B$ on the traveling train. Just when the flashes of lightning occur (as judged from the embankment), this point M' naturally coincides with the point M, but it moves towards the right in the diagram with the velocity v of the train. If an observer sitting in the position M' in the train did not possess this velocity, then he would remain permanently at M, and the light rays emitted by the flashes of lightning A and B would reach him simultaneously, *i.e.* they would meet just where he is situated.

Now in reality (considered with reference to the railway embankment) he is hastening towards the beam of light coming from B, whilst he is riding on ahead of the beam of light coming from A. Hence the observer will see the beam of light emitted from B earlier than he will see that emitted from A. Observers who take the railway train as their reference-body must therefore come to the conclusion that the lightning flash B took place earlier than the lightning flash A. We thus arrive at the important result:

Events which are simultaneous with reference to the embankment are not simultaneous with respect to the train, and *vice versa* (relativity of simultaneity). Every reference-body (co-ordinate system) has its own particular time; unless we are told the reference-body to which the statement of time refers, there is no meaning in a statement of the time of an event.

In his thought experiments, Einstein usually assumed that the space inside a train or chest is an independent coordinate system regardless of the relative motion of the train or chest with respect to the outside environment (as longs as the train or chest is in uniform motion).

Another important fact we should remember is that the premise (postulation) of relativity is that the speed of light is the same for any observer (or coordinate system) regardless of the state of motion of the observer/coordinate system.

It is unclear whether the long train (the length of the train is not important) in the Einstein's thought experiment is *airtight one* or *open canopy* (no canopy) style. So let us check the two different cases as in the following exercises:

Exercise 15-1

Let us assume that the train in **Fig. 15-1** is airtight style and the space inside the train is an inertial environment/ coordinate system. Even if the train is airtight style, the light rays from the two lightening strokes can enter into the train through window pane. *M* is the midpoint between *A* and *B* on the embankment. *M'* is the midpoint between *A'* and *B'* on the train. In **Fig. 15-1**, three points *A'*, *M'*, and *B'* on the train correspond with three points *A*, *M*, and *B* on the embankment respectively at the moment two lightening strokes occur at *A* and *B* respectively. An observer on the train rests at midpoint M'.

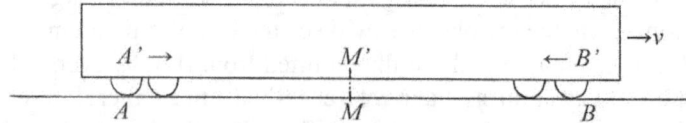

Fig. 15-2 Three points *A'*, *M'*, and *B'* on the train correspond with three points *A*, *M*, and *B* on the embankment respectively.

Let us assume again that the train starts moving at a uniform speed v in the direction shown by the arrow at the moment two lightening strokes occur at two points *A* and *B* simultaneously.

Question:

Will the observer at midpoint *M'* on the train see the two lights from *A* and *B* at the same time or different time?

Answer #1:

Since the train is an inertial coordinate system (according to Einstein's opinion) regardless of its motion with respect to the embankment, the speed of light from *A'* is the same as the speed of light from *B'* for the observer. Therefore, the observer will see the two lights at the same time. This means that Einstein's relativity of simultaneity is wrong.

On the other hand, my position is that the airtight train be an inertial coordinate system for sound or ordinary objects such as a wood block or iron ball. But we should think more of the matter whether the airtight train is also an inertial coordinate system for light. My position is that light speed is constant with respect to space (*absolute space*) and not with respect human observers or laboratory or train (see **Chapters 19-20**).

In addition, the motion of the train is not in inertial state because the earth, on which the train moves, is in complicated motion consists of rotation motion and revolution motion in space. Therefore any object or laboratory on the surface of the earth cannot be in inertial state. Even if readers agree with the idea that the speed of light is constant with respect to absolute space, we cannot not tell whether the light from the head-portion (B') of the train arrives at the observer at midpoint M' earlier than the light from the rear-portion (A') because we cannot tell in which direction the train moves with respect to absolute space.

I think the given situation is the similar to the ***Sagnac experiment***; the surface of the earth, on which the train moves, is an interferometer of the Sagnac experiment and the observer at the midpoint M' corresponds with the observing screen of the interferometer. Through his experiment Sagnac proved that the speed (relative speed) of light is not the same (see the Sagnac experiment at the latter part of this chapter).

Exercise 15-2

Let us assume that the embankment is stationary with respect to space (absolute space) and that the train has only platform and no canopy. In this case, the space and air above the platform of the train does move at the same speed as the train. In this case, our common sense will tell us that the light from the head-portion (B') arrives at the observer at midpoint M' earlier than the light from the rear-portion (A') of the train.

In this case (if the train is o open-canopy style), we do not need the cumbersome train at all. I mean we can simplify the experiment as shown in **Fig. 15-3**.

In **Fig. 15-3**, two lightening strokes occur at two places—A and B—on the embankment simultaneously. M is the mid-point between A and B. That is, $AM = MB = a$. Observer O starts moving from M toward B at the uniform speed v at the moment the two strokes of lightening occur at A and B simultaneously.

Harvard Physicists Confuse Relative Speed with Proper Speed

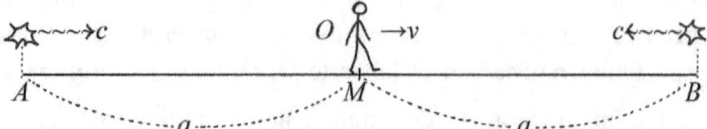

Fig. 15-3 Observer O starts moving from midpoint M toward B at the uniform speed v at the moment two strokes of lightening occur at A and B at the same time.

Questions:

Find the following from the Newton-Galilean point of view and then from the relativistic point of view.

(1) The time (t_1) the light ray from B takes to get to observer O.

(2) The time (t_2) the light from A takes to catch up with observer O.

(3) The relative speed between the observer and the light coming from B toward the observer.

(4) The relative speed between the observer and the light coming from A toward the observer.

Newton-Galilean solution:

Fig. 15-4 shows that observer O meets the light ray coming from B at time t_1.

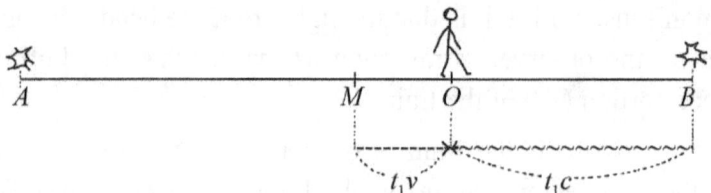

Fig. 15-4 Observer O meets the light from B at time t_1.

(1) In the diagram, $t_1 v + t_1 c = MB = a$.

Therefore, $t_1 = \dfrac{a}{c+v}$.

Special Relativity

(2) **Fig. 15-5** shows that the light ray coming from A catches up with observer O at time t_2.

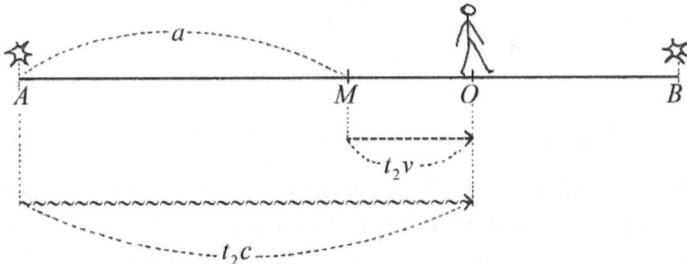

Fig. 15-5 Light from A catches up observer O at time t_2.

In the diagram, $t_2 c = a + t_2 v$. Therefore, $t_2 = \dfrac{a}{c-v}$.

Obviously, $t_1 \left(= \dfrac{a}{c+v}\right) < t_2 \left(= \dfrac{a}{c-v}\right)$. That is, observer O sees the light coming from B earlier than the light from A.

(3) The *relative speed* of the light from B with respect to observer O is $c + v$, and vice versa. "$c + v$," which is larger than c, is possible because this is a relative speed. Yet the proper speed of the light from B is c with respect to free space.

(4) The relative speed of the light from A with respect to observer O is $c - v$, and vice versa. Yet the proper speed of light is still c.

Relativistic solution:

According to the isotropy of light, the *relative speed* between the observer and the light ray from B toward the observer is $c + v = c$.

Therefore, in **Fig. 15-4**, $t_1 = \dfrac{a}{c+v} = \dfrac{a}{c}$.

The *relative speed* between observer O and the light from A is $c - v = c$.

Therefore, in **Fig. 15-5**, $t_2 = \dfrac{a}{c-v} = \dfrac{a}{c}$.

The above result means that Einstein's relativity of simultaneity is wrong. We do not consider length contraction in this case. If we do, we get the following result:

For observer O, distance $a\ (= MB)$ will be

$$a\sqrt{1-v^2/c^2} = a\sqrt{\frac{(c+v)(c-v)}{c^2}} = a\sqrt{\frac{c \times c}{c^2}} = a.$$

For light, distance a will be $a\sqrt{1-c^2/c^2} = 0$.

Exercises 15-1 and **15-2** are basically a kind of Sagnac experiment; the observer at the midpoint of the train is equivalent to a viewing screen, and the train is equivalent to an interferometer in motion. Sagnac proved, in 1913, that light speed (*relative speed*, not proper speed) does change. But Sagnac, as well as other physicists, did not distinguish proper speed from relative speed.

Sagnac Effect

Relativists interpret the Sagnac effect in terms of special relativity, and they argue that Sagnac effect proves the invariance of light speed. Many dissidents interpret Sagnac effect as the evidence of *anisotropy* of light speed and the evidence of the absoluteness of space. However, since both relativists and dissidents do not distinguish the concept of proper speed from that of relative speed they have limited understanding of Sagnac experiment.

Doug Marett (a dissident) correctly states that light speed varies and that space (universe) is the absolute frame of rest. [2] But he does not discern the concept of proper speed of light (which is constant with respect to absolute space) from that of relative speed of light (which is observer-dependent).

Marett says, "Although multiple stationary frames can be identified depending on the circumstances, there is only one truly unique stationary frame, and this is the frame of absolute space, a.k.a 'the fixed stars.'" Marett made a mistake; I have explained that the fixed stars cannot be the absolute frame of reference because of *the Fourth law of relative speed* (see **Chapter 6**). According to this law, the speed of any object on the earth with respect to a (any) fixed star is zero. I think that the speed of light is constant with respect to space. I explained that space is absolute without the presence of ether (see **Chapter 18**). I suggested the method of measuring absolute speed in **Chapters 21-22**.

Special Relativity

Concerning the relativity of simultaneity, Einstein supposes that he can place synchronized clocks in more than two places in his book Relativity (fifteenth ed.) as follows: [3]

> For this purpose we suppose that clocks of identical construction are placed at the positions A, B, and C of the railway line (co-ordinate system), and that they are set in such a manner that the positions of their pointers are simultaneously (in the above sense) the same. ....
>
> When two clocks arranged at rest in two different places of a reference-body are set in such a manner that a particular position of the pointers of the one clock is simultaneous (in the above sense) with the same position of the pointers of the other clock, then identical "setting" are always simultaneous (in the sense of the above definition).

In the above statement, Einstein acknowledges the concept of **universal time**. I agree with the idea that we can place synchronized clocks in more than two places. Then, we can fill the whole universe with synchronized clocks according to the method of Einstein, at least theoretically. Why not?

This means that Einstein acknowledged the ***absolute universal time*** as long as clocks are at rest. The remaining issue is whether a moving clock, say clock A, goes slower than clock B at rest when viewed from clock B. However, according to the reciprocity of relative motion/speed, it is also true, according to relativity, clock B goes lower than clock A *when viewed from* clock A.

Relativists say that we (they) can replace the two clocks with two persons or two insects or two whatever, and that the phenomenon of "aging less than the other" would happen as well. This is the story of the twin paradox (see **Chapter 29**). But sadly, the speed in Einstein's theory of relativity is proper speed (which is observer-free) and not relative speed (which is observer-dependent) (see **Chapter 7**).

Einstein relied on the speed of light too much. For him, the speed of light (c), which is neither relative speed nor proper speed, is the measure of all things. There is no such light speed in reality.

Whether one sees two lights from two different events at the same time or at different time has nothing to do with the judgment of the simultaneity of the two events. Why should we depend on light speed whose speed is limited? If the seed of light were infinite (∞), we would have no problem in judging the simultaneity of any two or more events. If God's speed of gathering information is infinite (∞), God must have no problem in judging the simultaneity of events.

But we can judge the simultaneity of events quite easily and correctly without asking almighty God; if we find the distances of the venues of the events and the speed of light, we can correctly judge whether the events occurred at the same time or not. I think that space and time are absolute and that **universal time** is real and **universal simultaneity** exists. ♦

16

Einstein' concept of Inertial motion Is Incorrect

Einstein's concept of inertial motion is incorrect. Einstein explained inertial motion in his book *Relativity* (16th ed.) as follows: [1] (Paragraph division is redone.)

> As is well known, the fundamental law of the mechanics of Galilei-Newton, which is known as the law of inertia, can be stated thus: A body removed sufficiently far from other bodies continues in a state of rest or of uniform motion in a straight line. This law not only says something about the motion of the bodies, but it also indicates the reference-bodies or system of co-ordinates, permissible in mechanics, which can be used in mechanical description. The visible fixed stars are bodies for which the law of inertia certainly holds to a high degree of approximation.
>
> Now we use a system of co-ordinates which is rigidly attached to the earth, then, relative to this system, every fixed star describes a circle of immense radius in the course of an astronomical day, a result which is opposed to the statement of the law of inertia. So that if we adhere to this law we must refer these motions only to systems of co-ordinates relative to which the fixed stars do not move in a circle. A system of co-ordinates of which the state of motion is such that the law of inertia holds relative to it is called a "Galileian system of co-ordinates." The laws of the mechanics of Galilei-Newton can be regarded as valid only for a Galileian system of co-ordinates.

Some dissenters, who believe in the mechanics of Galilei-Newton and the absoluteness of space and time, think that the fixed stars can be the reference body of the absolute motion of things in space. Einstein rejected such an idea for the reason that the positions of *fixed stars are not frozen but they describe a circle of immense radius in the course of an astronomical day*."

Both Galilei-Newtonian physicists and Einstein are wrong; here is the reason why: The phenomenon that each fixed star looks stationary (frozen in the sky) (due to the vast distance from the earth) means, conversely speaking, that the motion of the earth or the even our galaxy is frozen (zero) with respect to the fixed star.

Though the earth moves at a considerable speed in space (though the motion of the earth in space is neither linear nor uniform), the relative motion/speed of the earth with respect to a *fixed star* is **zero** [0] if the distance of the fixed star is literally infinite (∞). The motion of a fixed star and that of the earth are mutually frozen! This is simply proved by **Eq. (2)** and **Eq. (3)** as follows: (*r* or *r*-factor in the equations means the distance between an observer and the line of motion of an object being observed. *v* is the proper speed of the object. See **Chapter 5**.)

If $r = \infty$ in **Eq. (2)**,

$$\lim_{r \to \infty} \frac{v(vt - m)}{\sqrt{(vt - m)^2 + r^2}} = 0.$$

If $r = \infty$ in **Eq. (3)**,

$$\lim_{r \to \infty} \frac{v^2 t}{\sqrt{(vt)^2 + r^2}} = 0.$$

I refer to this phenomenon as ***the fourth law of relative speed***.

The Fourth Law of Relative Speed

If *r*-factor (see **Chapter 5**) is very large (∞), the relative speed between two involved objects is zero (0) even when the observation angle is not 90°.

The fourth law of relative speed explains the reason why fixed stars look frozen. But it does not mean that fixed stars can be the reference body for the ***absolute motion*** of things on the earth or the earth itself.

What about the Einstein's argument that "every fixed star describes a circle of immense radius every day" when viewed from the earth? This phenomenon is a ***virtual motion*** of the fixed stars due to the rotation motion of the earth. The fixed stars do not ***actually*** circle around the earth at the marvelous speed (perhaps infinite times the speed of light!).

How do we know that the circular motion of the fixed stars (with respect to the earth) is virtual motion whereas the rotation motion of the earth is actual motion? That is we check whether the centrifugal force occurs; if centrifugal force occurs due to the rotation motion of circular motion, the rotation or circular motion is actual one; if not the rotation or circular motion is virtual one. ***Coriolis Effect*** (***Force***) proves that the earth is in *actual* rotation motion (see the latter part of this chapter).

It is not that all fixed stars describe immense circle in the sky. The size (radius) of the circle (virtual circle) depends of the angle of a fixed star in relation with the axis of the rotation of the earth. The smaller the angle, the smaller circle the fixed star describes. This principle explains the phenomenon that the North Star (Polar Star) looks stayed put in the photos taken in the northern hemisphere (see **Fig. 6- 2** in **Chapter 6**).

Einstein did not distinguish actual motion from virtual motion. Reciprocity holds in relative motion, which is either actual or virtual. But reciprocity does not hold in actual motion. The motion of an fixed star (or any heavenly object) with respect to space (absolute space) is actual and absolute motion. Einstein was wrong because he thought that all motion is relative and reciprocal. Einstein wrongly asserted that the state of being accelerated is the same as the state of being at rest as follows: [2] (The words in brackets are mine.)

> Even though it [a chest] is being accelerated with respect to the 'Galilean space [translated as free space]' first considered we can nevertheless regard the chest as being at rest. We have thus good grounds for extending the principle of relativity to include bodies of reference which are accelerated with respect to each other.

I do not agree with Einstein's *relativity of acceleration* (see **Chapter 34**). Einstein also wrongly regarded a rotating disc as being at rest as follows: [3]

> An observer who is sitting eccentrically on the disc K' is sensible of a force which acts outwards in a radial direction, and

> which would be interpreted as an effect of inertia (centrifugal force) by an observer who was at rest with respect to the original reference-body *K*. But the observer on the disc may regard his disc as a reference-body which is "at rest"; on the basis of the general principle of relativity he is justified in doing this.

Rotation motion is a kind of accelerated motion. Centrifugal force, which Einstein misidentified with gravity, occurs only when an object is in actual rotation/circular motion and not when the object is in virtual rotation/circular motion.

Inertial motion is constant in speed and direction *with respect to absolute space*. That is, inertial motion is absolute motion and actual motion. (Note that actual motion is not necessarily absolute motion; for example, the actual motion of a car on the surface of the earth is not absolute motion.) Absolute motion and actual motion is not relative motion.

Einstein's idea that any object or observer can be inertial frame (coordinate system) of reference is wrong. For example, the earth is not in an inertial sate because of its complicated motion (consists of rotation motion and circular motion) in space. Therefore, any object or laboratory on the earth is not and cannot be an inertial frame of reference. The experiment of Newton's law of motion on the laboratory on the earth is approximately correct. Objects on the surface of the earth has some size of centrifugal force (inertial force) dud to the rotation motion of the earth; *Coriolis force* or *Coriolis Effect* occurs due to the centrifugal force (inertial force) of the rotation motion of the earth.

Coriolis Effect

(Source: http://en.wikipedia.org/wiki/Coriolis_effect)

In physics, the Coriolis Effect is a deflection of moving objects when they are viewed in a rotating reference frame. In a reference frame with clockwise rotation, the deflection is to the left of the motion of the object; in one with counter-clockwise rotation, the deflection is to the right. Although recognized previously by others, the mathematical expression for the Coriolis force appeared in an 1835 paper by French scientist Gaspard-Gustave Coriolis, in connection with the theory of water wheels. Early in the 20th century, the term Coriolis force began to

Special Relativity

> be used in connection with meteorology.
>
> Newton's laws of motion describe the motion of an object in a (non-accelerating) inertial frame of reference. When Newton's laws are transformed to a uniformly rotating frame of reference, the Coriolis and centrifugal forces appear. Both forces are proportional to the mass of the object. The Coriolis force is proportional to the rotation rate and the centrifugal force is proportional to its square. The Coriolis force acts in a direction perpendicular to the rotation axis and to the velocity of the body in the rotating frame and is proportional to the object's speed in the rotating frame. The centrifugal force acts outwards in the radial direction and is proportional to the distance of the body from the axis of the rotating frame. These additional forces are termed inertial forces, fictitious forces or pseudo forces. They allow the application of Newton's laws to a rotating system. They are correction factors that do not exist in a non-accelerating or inertial reference frame.

Einstein's concept of inertial motion or inertial frame of reference is incorrect.

I suggested a method of determining the magnitude of absolute motion of things based on the idea that the speed (proper speed) of light is constant with respect to absolute space (see **Chapter 22**). ♦

17

The Motion of Photons Is Different from That of Ordinary Objects

Unlike ordinary objects such as a car, train, airplane, spacecraft, etc., photons (light) have no inertia. Light or electromagnetic waves are neither accelerated nor decelerated by the motion of third parties (objects or observers). Ordinary objects are subject to the Newton's laws of motion, but electromagnetic waves are not. The behavior of charged particles/objects is different from that of non-charged ones. The argument that special relativity united electrodynamics, optics, and classical mechanics into one is only **mathematical delusion**.

According to relativists, the speed of a photon (light) is always the same (c) relative to all observers, each of which moves truly in arbitrary and unpredictable fashion. This means that the speed of light is *extremely flexible* and thus *not constant*. Logically, if a *variable* (the velocity of an observer) plus an unknown (light speed) = constant (c), the unknown (light speed) must be a variable.

In order to explain (rationalize) such the *isotropy* of light speed, an **improbable assumption**, Einstein introduced **space-time continuum** whose supposed role is to expand or contract or curve in order to make the speed of a photon (light) always the same (c) for any/all observers/objects. This is even more improbable an assumption. Does space-time have any sense or omniscient ability to react to the motion of all the photons, observers, and objects in the world to make the speed of light the same all the time?

Relativists think that the Lorentz transformation (LT) or the formula of speed addition proves the constancy of light speed. But the LT and the

Special Relativity

speed addition formula were formulated in order to make the light speed of light always the same for all observers/objects. Therefore, the attempt to prove the constancy of light speed by the LT or the speed addition formula is but a circular reasoning or vicious circle.

The postulate (assumption) that the speed of light is always the same for any observer regardless of whether the observer moves in the direction of the light or in the opposite direction of the light ($c - v = c$, and $c + v = c$) is obviously unphysical. This postulate came from not distinguishing proper speed from relative speed. Einstein mistook proper speed, which is observer-free, for relative speed, which is observer-dependent.

When it comes to proper speed, light speed might be the upper limit of speeds. But when it comes to relative speed, a photon can have a relative speed of up to two times the light speed ($2c$) with respect to a counterpart object. For example, if two photons move in the opposite directions of each other on the same straight line, the **relative speed** between the two is $2c$. If the two photons move in the same direction on the same straight line, the relative speed between the two is zero (0). If an observer and a photon move in the opposite direction on the same straight line at the proper speeds of v and c respectively, the relative speed of the observer with respect to the photon (or) is $v + c$, and *vice versa*. If the observer and the photon move in the same direction on the same straight line at the proper speeds of v and c respectively, the relative speed between the two is $v \sim c$. If a photon and an observer are in a 2-D or 3-D situation, the relative speed between the two changes all the time. Yet the proper speed of light is always the same (c) in any of the above situations.

In Lorentz's time, physicists, including Lorentz himself, did not distinguish the concept of proper speed from that of relative speed. Lorentz and Einstein and all other relativists, including Harvard physicists mistook proper speed for relative speed. Einstein's concept of speed was limited to 1-D situation (see **Chapter 24**). This is the reason that the formula of the addition of speeds is applicable only to 1-D situations.

In a word, the postulate of the *isotropy of light speed* reveals the advocates' (relativists') illiteracy in physical science. ♦

18

Are Space and Time Relative or Absolute?

Einstein had published special relativity (1905) and general relativity (1915) based on the idea that space and time are relative. Relativists think that all physical quantities—such as distance, speed (except the speed of light), time, mass, energy, gravity, etc.—are relative. Hence the term *relativity*. However, in 1920 (1922 in some other sources), Einstein publicly denied the idea that space and time are relative in his public address at Leiden University (Germany). Thus Einstein recognized the absoluteness of space and time. Einstein says in his Leiden Address as follows: [1]

> To deny the ether is ultimately to assume that empty space has no physical qualities whatever. The fundamental facts of mechanics do not harmonize with this view.... Besides observable objects, another thing, which is not perceptible, must be looked upon as real, to enable acceleration or rotation to be looked upon as something real....The conception of the ether has again acquired an intelligible content, although this content differs widely from that of the ether of the mechanical wave theory of light....
>
> According to the general theory of relativity, space is endowed with physical qualities: in this sense, there exists an ether. Space without ether is unthinkable; for in such space there not only would be no propagation of light, but also no possibility of existence for standards of space and time (measuring-rods and clocks), nor therefore any space-time intervals in the physical sense. But this ether may not be thought of as endowed with the qualities of ponderable media, as consisting of parts which may be tracked through time.

Special Relativity

Einstein tries to explain that his *ether* is different sort from what physicists of the 19th and early 20th centuries believed to be. But the difference is not clear. Einstein still thinks his ether as the medium through which light propagates. It is interesting that Einstein says, *"Without ether there would be no possibility of existence for standards of space and time.* Relativity was the theory that there are no standards of space/distance and time. Einstein's acknowledgement of ether or the standard of space and time means that he gave up both special and general relativity.

Einstein recognized the absoluteness of space in ruminating over his disc thought experiment from which he had derived general relativity (see **Chapter 40**). Einstein had formulated general relativity based on the idea that centrifugal force (= inertial force) is the same as gravity (see **Chapters 36** and **40**). But later, he realized that he could not explain centrifugal force without recognizing the absoluteness of space. He realized that if space were relative there would be no physical difference between when a disc rotates and when the disc does not. Centrifugal force, which Einstein identified with gravity, occurs only when a disc rates and not when the disc does not rotate. Therefore, Einstein could not but recognize the absoluteness of space.

The experiment known as ***Newton's bucket*** is interpreted by classical physicists as the proof of the absoluteness of space. Einstein denied the interpretation of classical mechanists. The experiment of Newton's bucket is interesting and important enough to investigate. Walter Isaacson introduces the controversy over the phenomenon of Newton's bucket as follows: [2] (Reference numbers in the citation are reset for this book. **Fig.19-1** is my addition.)

> Einstein and Besso also looked at whether rotation could be considered a form of relative motion under the equations of the *Entwurf* theory. In other words, imagine that an observer is rotating and thus experiencing inertia. Is it possible that this is yet another case of relative motion and is indistinguishable from a case where the observer is at rest and the rest of the universe is rotating around him?
>
> The most famous thought experiment along these lines was that described by Newton in the third book of his *Principia*.

Imagine a bucket that begins to rotate as it hangs from a rope. At first the water in the bucket stays rather still and flat. But soon the friction from the bucket causes the water to spin around with it, and it assumes a concave shape. Why? Because inertia causes the spinning water to push outward, and therefore it pushes up the side of the bucket.

Yes, but if we suspect that all motion is relative, we ask: What is the water spinning relative to? Not the bucket, because the water is concave even when it is spinning along with the bucket, and also when the bucket stops and the water keeps spinning inside for a while. Perhaps the water is spinning relative to nearby bodies such as the earth that exert gravitational force.

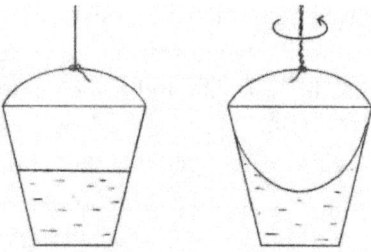

Fig. 18-1 Newton's Bucket

But imagine the bucket spinning in deep space with no gravity and no reference points. Or imagine it spinning alone in an otherwise empty universe. Would there still be inertia? Newton believed so, and said it was because the bucket was spinning relative to absolute space.

When Einstein's early hero Ernst Mach came along in the mid-nineteen century, he debunked this notion of absolute space and argued that the inertia existed because the water was spinning relative to the rest of the matter in the universe. Indeed, the same effects would be observed if the bucket was still and the rest of the universe was rotating, he said. [3] The general theory of relativity, Einstein hoped would have what he dubbed "Mach's Principle" as one of its cornerstones. Happily, when he analyzed the equation in his *Entwurf* theory, he concluded that they *did* seem to predict that the effects would be the same whether a bucket was spinning or was motionless while the rest of the universe spun around it.

Or so Einstein thought. He and Besso made a series of very clever calculations designing to see if indeed this was the case. In

Special Relativity

their notebook, Einstein wrote a joyous little exclamation at what appeared to be the successful conclusion of these calculations: "Is correct."

Unfortunately, he and Besso had made some mistakes in this work. Einstein would eventually discover those errors two years later and realize, unhappily, that the *Entwurf* did not in fact satisfy Mach's principle. In all likelihood, Besso already warned him that this might be the case. In a memo that he apparently wrote in August 1913, Besso suggested that a "rotation metric" was not in fact a solution permitted by the field equations in the *Entwurf*. (pp. 199-201)

The *Entwurf* theory or simply the *Entfurf* was a mathematical study of Einstein and Grossmann (mathematician) in 1913. [4] It was a tentative and unsuccessful theory of general relativity (see **Chapter 42**).

My opinion about the Newton's bucket effect is that the rise of water in the bucket is cause by both inertia and earth gravity. If there were no gravity, the rising of water would not occur. We can check this readily by a simple experiment in spaceship. In this regard, Newton was half correct and half wrong. Mach and Einstein were wrong in the point they said that "the effects would be the same whether a bucket was spinning or was motionless while the rest of the universe spun around it." We cannot let the buck be motionless while we let the universe spin around it. But it is easy to reason that centrifugal force (inertial force) occurs only while an object is in rotation motion relative to the universe (absolute space) and not the way around.

The absoluteness of space and time is supported by inertial motion/force and the ***gyroscopic inertia*** (see **Chapter 19**). Later, in 1920, Einstein acknowledged the absoluteness of space in his address at Leiden because he realized that he could not explain the inertial force which occurred in his disc thought experiment (see **Chapter 40**).

By recognizing the absoluteness of space and time, Einstein destroyed, though he did not realized it, the foundation of his theory of relativity. Einstein was not aware of what his acknowledgement of the ether (the absoluteness of space) had to do with his theory of relativity. Strangely enough, after he acknowledged the absoluteness of space and time, Einstein kept his original idea that space-time is relative and that space is movable and expandable. Einstein states in the "Note to the Fifteenth Edition" of his book *Relativity* (fifteenth ed.) as follows: [5]

> In this edition I have added, a fifth appendix, a presentation of my views on the problem of space in general and on the gradual modifications of our ideas on space resulting from the influence of the relativistic view-point. I wished to show that space-time is not necessarily something to which one can ascribe a separate existence, independently of the actual objects of physical reality. Physical objects are not in space, but these objects are spatially extended. In this way the concept of "empty space" loses its meaning.

Einstein's concept of space is that of Descartes (1596-1650). The *Appendix Five* of his fifteenth edition of *Relativity* (1952) is as follows: [6]

Relativity and the Problem of Space

It is characteristic of Newtonian Physics that it has to ascribe independent and real existence to space and time as well as to matter, for in Newton's law of motion the idea of acceleration appears. But in this story, acceleration can only denote "acceleration with respect to space." Newton's space must thus be thought of as "at rest," or at least as "unaccelerated," in order that one can consider the acceleration, which appears in the law of motion, as being a magnitude with any meaning. Much the same holds with time, which of course likewise enters into the concept of acceleration. Newton himself and his most critical contemporaries felt it to be disturbing that one had to ascribe physical realities both to space itself as well as to its state of motion; but there was at that time no other alternative, if one wished to ascribe to mechanics a clear meaning.

It is indeed an exacting requirement to have to ascribe physical reality to space in general, and especially to empty space. Time and again since remotest times philosophers have resisted such a presumption. Descartes argued somewhat on these lines: space is identical with extension, but extension is connected with bodies; thus there is no space without bodies and hence no empty space. The weakness of this argument lies primarily in what follows. It is certainly true that the concept extension owes its origin to our experiences of laying out or bringing into contact solid bodies. But from this it cannot be concluded that the concept of extension may not be justified in cases which have not themselves given rise to the formation of this concept. Such an enlargement of concepts can be justified indirectly by its value for the comprehension of empirical results. The assertion that extension is confined to bodies is therefore of itself certainly unfounded. We shall see later, however, that the general theory of relativity confirms Descartes' conception in a roundabout way. What brought Descartes to his remarkably attractive view was certainly the feeling that, without compelling necessity, one ought not to ascribe reality to a thing like space, which is not capable of being "directly experienced."

The psychological origin of the idea of space, or of the necessity for it, is far from being so obvious as it may appear to be on the basis of our customary habit of thought. The old geometers deal with conceptual objects (straight line, point, surface), but not really with space as such, as was done later in analytical geometry. The idea of space, however, is suggested by certain primitive experiences. Suppose that a box has been constructed. Objects can be arranged in a certain way inside the box, so that it becomes full. The possibility of such arrangements is a property of the material object "box," something that is given with the box, the "space enclosed" by the box. This is something which is different for different boxes, something that is thought quite naturally at being independent of whether or not, at any moment, there are any objects at all in the box. When there are no objects in the box, its space appears to be "empty."

So far, our concept of space has been associated with the box. It turns out, however, that the storage possibilities that make up this box-space are independent of the thickness of the walls of the box. Cannot this thickness be reduced to zero, without the "space" being lost as a result? The naturalness of such a limiting process is obvious, and now there remains for our thought the space without the box, a self-evident thing, yet it appears to be so unreal if we forget the origin of this concept. One can understand that it was repugnant to Descartes to consider space as independent of material objects, a thing that might exist without matter. (At the same time, this does not prevent him from treating space as a fundamental concept in his analytical geometry.) The drawing of attention to the vacuum in a mercury barometer has certainly disarmed the last of the Cartesians. But it is not to be denied that, even at this primitive stage, something unsatisfactory clings to the concept of space, or to space thought of as an independent real thing.

The way in which bodies can be packed into space (*e.g.* the box) are the subject of three-dimensional Euclidian geometry, whose axiomatic structure readily deceives us into forgetting that it refers to realisable situations.

If now the concept of space is formed in the manner outlined above, and following on from experience about the "filling" of the box, then this space is primarily a *bounded* space. This limitation does not appear to be essential, however, for apparently a larger box can always be introduced to enclose the smaller one. In this way space appears as something unbounded.

I shall not consider here how the concepts of the three-dimensional and the Euclidian nature of space can be traced back to relatively primitive experiences. Rather, I shall consider first of all from other points of view the role of the concept of space in the development of physical thought.

When a smaller box s is situated, relatively at rest, inside the hollow space of a larger box S, then the hollow space of s is a part of the hollow space of S, and the same "space," which contains both of them, belongs to each of the boxes. When s is in motion with respect to S, however, the concept is less simple. One is then inclined to think that s encloses always the same space, but a variable part of the space S. It then becomes necessary to apportion to each box its particular space, not thought of as bounded, and to assume that these two spaces are in motion with respect to each other.

Before one has become aware of this complication, space appears as an unbounded medium or container in which material objects swim around. But it must now be remembered that there is an infinite number of spaces, which are in motion with respect to each other. The concept of space as something existing objectively and independent of things belongs to pre-scientific thought, but not so the idea of the existence of an infinite number of spaces in motion relatively to each other. This latter idea is indeed logically unavoidable, but is far from having played a considerable role even in scientific thought. --------

The subtlety of the concept of space was enhanced by the discovery that there exist no completely rigid bodies. All bodies are elastically deformable and alter in volume with the change in the temperature. The structures, whose possible congruences are to be described by Euclidian geometry, cannot therefore be presented apart from physical concepts. But since physics after all must make use of geometry in the establishment of its concepts, the empirical content of geometry can be stated and tested only in framework of the whole of physics.

Einstein's concept of space is vague and inconsistent; his concept of space was sometimes Cartesian or otherwise at other times. In the above citation, Einstein thinks that the space inside the smaller box s is independent and freely movable with respect to the space inside the larger box S. Einstein says that there is infinite number of spaces in motion relative to each other. This idea is starkly against Cartesian concept of space. Can the space confined by a box, whose wall has no thickness, move around freely? Einstein's concept of space is bizarre and unrealistic. Space has only the capacity to contain any objects and the space itself is not material. I do not think such space can be freely movable or curved or distorted by the motion of objects or the presence of gravity.

I think space is neither movable nor deformable and the dimension/volume of a (any) portion of space is absolute. I, for one, thought the idea that space is absolute and immovable is much more comfortable and rational than the idea that space is freely movable and deformable. My idea is this: If light is to cover the distance between two given points that are at rest in absolute space, it takes the light the finite or absolute time, so that the distance between the two points divided by the speed (proper speed) of light (c) is constant and absolute. In other words, the proper speed of light with respect to absolute space is constant.

Einstein thought that the speed of a photon (light) is the same for human observers or any other objects (at the same time!). In this respect, Einstein concept of light speed has nothing to do with space. Einstein's idea means that the light speed varies depending on the motion/speed of objects (including humans). If not, how can the speed of light constant with respect to any observer/object in motion?

The concept of uniform linear motion, inertial motion, inertial force, gyroscopic inertia, etc. support the absoluteness of space and time (see **Chapters 18-19**). If space and time are relative as Einstein thought, we cannot define or determine the uniform motion, inertial motion, acceleration, inertial force, etc. The reason Einstein reintroduced *an ether* in his address at Leiden was because he realized that there would be no standard of distance or speed if space and time were relative entities.

Many dissidents recognize the problem with relativistic concept of space, and they agree with the necessity of *the absoluteness of space and time*. Therefore, many dissidents revive the concept of *material ether* to explain the absoluteness of space or the propagation of light. But I think photons do not need material ether of any sort medium to propagate through space. This idea resembles the Newton's **emission theory** or **corpuscular theory** of light. [7] I do not mean light is made of particles. Light has properties of both particles and waves.

Space looks empty, but it is **not completely nothing**. Space is absolute in the point that it has absolute and immovable character. Some philosophers or scientists think that space does not exist as an independent entity. I think space is independent of the existence of materials, energy, or *fields* (such as gravitational field, electromagnetic field, etc.)

Let us think of a vacant space between two buildings. There is nothing between the two buildings except vacant space. But the vacant space between the two buildings has a certain dimension. This dimension is neither collapsible nor expandable (unless any catastrophic happenings occur to the surface of the earth). It has a certain capacity to hold certain amount of material or objects—houses, buildings or other facilities.

The vacant space in the above story is on the surface of the earth—a planet in motion in space. So the vacant space between two buildings on the surface of the earth is different from the absolute space of universe. I think that the whole universe minus all kinds of matters or material objects, energy, and fields in it remains unchanged in dimension and position no matter what kind of heavenly bodies move around in it. But I do not mean the space or universe has a limited dimension. I do not know whether the dimension of universe is finite or infinite. But I think that a portion or the whole space does not change in the dimension or position by any means. I think that time is also absolute and independent an entity and that the *rate of time flow* is not affected by any motion of an object or matter or gravity.

Einstein thought that space and time do not exist as independent

Special Relativity

entities but they form a unity called space-time continuum. The mission of the unity (space-time) is to make the speed of light always the same for all observers or objects, according to Einstein. (Relativists do not care if the light speed is proper speed or relative speed.) Einstein's opinion is that the relative speed of a photon is always the same (c) for all the objects in the universe including other photons, stars, Harvard physicists, etc. (whose motion, speed, and direction are truly arbitrary and unpredictable). If the term *light speed* is *proper speed*, then relativists' argument is true. But what Einstein and his followers mean is that the *relative speed* of a photon with respect to each and every object in the whole universe is always the same (c). I think this is more than impossible. I do not think space and time have such ability or responsibility to obey the law set by a physicist who had no ability to discern relative speed from proper speed.

Einstein seemed to think that the space occupied by an object is increased if the volume of an object is enlarged by the rise of temperature. Einstein identified the volume of an object with the volume of space occupied by the object. This is a Cartesian way of thinking. We can increase or decrease the volume of a balloon, but this does not mean that the space occupied by the balloon is increased or decreased; we can only change the volume of gas inside the balloon. The phenomenon of bomb explosion is that the gas and debris scatter in space due to the sudden increase of volume of the gas. But it does not mean that space itself is expanded.

Relativists believe in the **Big Bang** theory and the expansion of space/ universe. Local explosion of stars might be possible. But I do not think that the whole universe came into being suddenly by the Big Bang and that the universe or pace is still expanding at faster-than-light speed at the out front of the universe. Relativists say that the rate of expansion of space at the Hubble sphere is the same as the speed of light, and the rate of expansion beyond the Hubble sphere is faster than the speed of light. Relativists say that this is not contradictory with the upper limit of speeds because the faster-than-the-light speeds of heavenly bodies beyond the Hubble Sphere is due to the expansion of space itself not the motion of the heavenly bodies. This is a poor and unfounded explanation. With so many fallacies and contradictions of relativity, relativists have little to defend the Big Bang theory. The Big Bang theory is the 20th century version of Genesis that has nothing to do with science. ♦

19

The Gyroscope and Absolute Space

The World Book Encyclopedia states that gyroscopic inertia is the ability of the spinning axle of a gyroscope always to point in the same direction, no matter how the support of the gyroscope moves about. [1]

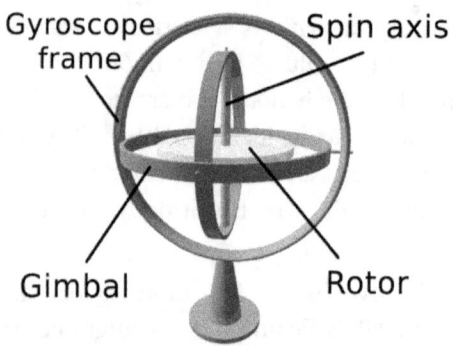

Fig. 19-1 The structure of a gyroscope
(Source: http://en.wikipedia.org/wiki/Gyroscope)

Gyroscopic inertia supports the absoluteness of space. If a gyroscope spins rapidly about its axis, the orientation (direction) of the axis *remains fixed*. Remains fixed *relative* to **what**? The answer is **absolute space**. In other words, the orientation of the axis is constant with respect to **absolute space**.

I interpret that *gyroscopic inertia* has to do with Newton's first law of motion (law of inertia). Any *matter point* that sits eccentrically on a spinning disc (rotor) tends to move in the direction of the tangent of the "matter point" This is centrifugal force. This is due to the Newton's law of inertia. (Centrifugal force is inertial force).

Of course the position of a mater point and the direction of the tangent line change continuously since the disc rotate. But the tangent line of the matter point tends to remain in "the same *plane*" to which the axis of the gyroscope is perpendicular. The orientation of the "*plane*" is "fixed" with respect to absolute space because the tangent lines which constitute the plane are the lines of inertial force of matter points. Since the orientation of the *plane* is fixed, so does the direction of the axis of gyroscope with respect to **absolute space**).

If space is relative, as relativists believe it to be, there would be no difference whether or not a disc (or other object) rotates; there would be no reason for the phenomenon that gyroscopic inertia or centrifugal force occurs only when the disc rotates.

In free space (where there is neither matter nor gravity nor any cause of friction) the motion of an object is uniform and *rectilinear* unless any force is added or subtracted. This is Newton's law of inertia. Relative to **what** is the inertial motion *uniform* and *rectilinear*? The answer is **absolute space** and **absolute time**. If time and space are not absolute, there is no way to tell that inertial motion is uniform in speed and direction.

The state of *being* stationary with respect to absolute space is also a kind of *uniform* motion. In this case, the magnitude of absolute speed is zero (0). Even if an objects is in uniform motion in space (this means that the proper speed or actual speed is constant with respect to absolute space), the relative speed of the object with respect to a specific observer or counterpart object does vary depending on the state of motion of the latter (the observer or counterpart object). But what we are talking about now is the absolute motion/speed of the objects with respect to absolute space.

Relativists cannot define or explain inertial force or uniform motion because they (relativists) believe in the relativity of space and time. Einstein explains inertial motion vaguely and incorrectly in his book *Relativity* (fifteenth ed.) as follows: [2]

> As is well known, the fundamental law of the mechanics of Galilei-Newton, which is known as the law of inertia, can be stated

thus: A body removed sufficiently far from other bodies continues in a state of rest or of uniform motion in a straight line. .... The visible fixed stars are bodies for which the law of inertia certainly holds to a high degree of approximation.

It is surprising that Einstein recognized (?) the concept of inertial law and tried to explain inertial motion in terms of fixed stars. According to relativity, inertial motion or law of inertia is nonsense or meaningless. In any event, Einstein vaguely or partially recognized the concept of inertial motion or the law of inertia. But fixed stars are not the proper thing for the explanation of inertial motion. Fixed stars are not that fixed; they move however slow they look when observed from the earth.

One thing we should note is that the relative speed of the fixed star with respect to the earth (or observers on the earth) is zero (0) or almost zero in most cases due to *the fourth law of relative speed* (see **Chapter 6**). Therefore, fixed stars cannot be the standard with which we can define or determine the inertial motion/speed of an object on the earth.

If we do not recognize *the absoluteness* of space and time, we cannot explain the gyroscopic inertia and inertial motion (including centrifugal force) in general because of uniform motion or inertial motion has to do with absolute space and absolute time. If either time or space or both are relative entities, inertial motion or uniform motion/speed itself is not determined.

The invariance of light speed (proper speed) has to do with absolute space and absolute time. If space and time are relative entities, the invariance of light speed (proper speed) does not hold. In order to rationalize the invariance of light speed, relativists wrote the theory or myth that space-time contracts or expands or curves. The relativistic statement that *the speed of light is the same for all observers or inertial reference bodies* leads us to absolutely unphysical or nonsensical world. The unphysical stories such as time dilation, length/distance contraction, and mass increase came into being only for the rationalization of the invariance of light speed.

The premise of relativity is that all motions or speeds are relative. On the other hand, relativists also argue that the speed of light is absolute. Relativity started from self-contradictory or incompatible postulations as follows:

Special Relativity

- All motion/speed is relative.
- The motion/speed of photon (light) is absolute.

Gyroscopic inertia, the law of inertia, uniform motion/speed, inertial mass, accelerated motion, centrifugal force, etc. are not compatible with relativity. ♦

20

The Mistakes of Galileo

Galileo (1564 -1642), the father of classical relativity, explained relative motion for the first time. He said that it is impossible to tell the difference between a moving object and a stationary one without some outside reference to compare it against. [1]

Yet Galileo made the mistake of not distinguishing proper speed from relative speed. Though he did not negate the absoluteness of space and time, he was not seriously aware of the absoluteness of space and time. He did not pay attention to the fact that the uniformity of inertial motion has to do with absolute space and absolute time.

The reason Galileo did not pay attention to the absoluteness of space and time was probably he did not know of the special property of gyroscope. Gyroscope was invented in early 1800's, long after Galileo had died. *The World Book Encyclopedia* states the history of gyroscope as follows: [2]

> The first written record of a gyroscope is in Gilbert's *Annalen*, in 1818. This report described a gyroscope made in 1810 by a German, G. C. Bohnenberger. Jean Foucault, a French physicist, built a gyroscope in 1852 with which he demonstrated that the earth rotates on its axis. Foucault named the instrument a *gyroscope*, taking it from two Greek words, *gyros*, meaning *revolution*, and *skopein*, meaning to *view*. Thus, gyroscope means *to view the revolution of the earth*.

It was Galileo who first declared that all motion is relative (in

Special Relativity

classical sense). But he did not know the concept of proper speed which is independent of observers. Galileo did not distinguish proper speed from relative speed. This was the mistakes of Galileo. The word that all motion is relative is therefore untrue and misleading.

In fact, all scientists, including Galileo, Newton, Einstein, and all modern physicists, are not aware of the difference between the concept of proper speed and that of relative speed. Without the knowledge of proper speed, the knowledge of relative speed is incomplete and incorrect. Galileo did not know that the light is the same (invariant) with respect to absolute space since he had no knowledge about the absoluteness of space.

Galileo gave other scientists the wrong idea that all motion is relative. Relative motion does exist. But he was not aware of the concept of proper motion/speed or actual motion/speed with respect to a given inertial environment (see **Chapter 1**) which is independent of the motion of observers. Galileo also made the mistake of considering relative motion/speed only in one-dimensional situation; he did not consider the relative motion/speed in 2-D or 3-D situations.

Galileo negated neither absolute space nor absolute time. But he did not pay serious attention to the absoluteness of space and time. Galileo's principle of relativity (in classical sense) influenced Newton (1642-1727) and later scientists including Einstein. Though Newton understood the concept of Galileo's relativity, he, unlike Galileo, maintained his clear idea that space and time are absolute.

Galileo's idea of classical relativity (without the notion of absolute space and absolute time) helped Einstein invent relativity. Einstein thought that all motion is relative and that there is no absolute space and absolute time. Absolute space and absolute motion exist, and we can measure the absolute motion/speed with respect to absolute space. ♦

21

Relative to what Is Light Speed Constant?

The question "Relative to what is the speed of light constant?" is crucial in judging the verity of relativity. The second postulate of special relativity is that the speed of light is the same (c) for all observers (regardless of the state of motion of the observers).

Two Postulates of the Special Theory of Relativity

The first postulate: The laws of physics are the same in all inertial frames of reference.

The second postulate: The speed of light in free space has the same value in all inertial frames of reference.

The second postulate means that the speed (relative speed) of a photon is the same (c) for all objects (including human observers) in the whole world at the same time regardless of all the different speeds and directions of the objects. If this is true, a photon or light is not a natural entity but a supernatural one and relativity is no longer physical science but metaphysics or paranormal.

The second postulate came from not distinguishing the concept of proper speed, which is observer-free, from that of relative speed, which is observer-dependent. Proper speed has to do with environmental reference body whereas relative speed has to do with point reference body. (See the definitions of *environmental reference body* and *point reference body* in **Chapters 1** and **2**. Readers cannot understand this chapter if they have no clear knowledge of environmental reference body and point reference body.)

I reasoned, after I checked many misconceptions in the theory of relativity, that the proper speed of light is constant (c) with respect to absolute space and that the relative speed of light with respect to counter object (including other light) can be from zero [0] to $2c$—two times the proper speed of light.

Throughout this book I disproved relativity, and I proved that space and time are absolute (see **Chapters 18-19**) and that the speed (proper speed) of light is constant with respect to absolute space.

I think space as the primary inertial environment. If a chest or spacecraft is in inertial motion in space, the inside room of the chest or space craft can be called the secondary inertial environment. Of course there may be tertiary or quaternary inertial environment. One thing we should note is that space itself is neither movable nor deformable by any physical means. (Einstein thought that space is freely movable or deformable. See **Chapter 18**). Since proper speed of light is constant with respect to absolute space and space is not movable, it is not that the proper speed of light is the same for secondary or tertiary inertial environment. That is, the proper speed of light is not the same for all inertial environments. This has to do with the fact that photons do not have inertia.

When it comes to ordinary objects, such as a car, airplane, spacecraft, etc., the laws of inertia hold in all inertial environments. Newton's laws of motion hold in any kind (primary, secondary, tertiary, etc.) of inertial frames. In other words, the laws of Newton are the same in all inertial frames when it comes to ordinary objects. But when it comes to light whose proper speed is c only with respect to the primary inertial frame (space), the isotropy of light speed does not hold in the secondary or tertiary inertial frames.

These stories means that the two postulates of special relativity are not true, and this means that special relativity and general relativity (a derivative of special relativity) are wrong. Relativists "religiously believe" that space-time is freely movable and limitlessly stretchable and that the speed of light is the same in all frames or reference. Relativists are wrong. They even do not distinguish proper speed from relative speed.

Here is quite sad news for relativists: On June 30, 2013, I happened to read a very important paper entitled "A Review of One-Way and Two-Way Experiments to Test the Isotropy of the Speed of Light." [1] The authors are Md. Farid Ahmed et al of York University, Ontario, Canada.

This paper seemed to have been published in 2011 or earlier because the authors say that this paper was accepted in the Indian Journal of Physics on March 30, 2011. I deem this paper is an outstanding review of the most advanced experiments concerning the measurement of light speed.

The authors of this paper are relativists. But the outcome of the experiments dismayed the authors. The results (data) found by the highly accurate methods of one-way determination of light speed are against the expectation of relativists: The one-way light speeds observed were inconsistent whereas the two-way average light speed is quite accurate and consistent. Two-way (round-trip) averaged speed of light has little meaning for relativists who believe in the invariance of light speed.

The authors of the paper say, subduing their disappointment, "We can only establish the special relativity for the round-trip averaged speed of light." [1] I think this statement is the acknowledgement of the defeat of relativists. Relativity theory cannot be established on the constancy of the two-way (round-trip) averaged light speed. The disappointing outcomes of the most advanced methods of measuring the one-way light speed might be a strong sign of the total failure of the theory of relativity. This gloomy news for relativists must have been known to many relativists and dissenters. But it is strange that both sides--relativists and dissenters—are quite quiet about this very important news.

Following are the abstract, introduction, and discussion parts of the paper. The reference number is of the original. [1] (Paragraph divisions are redone by Ahn for visual tidiness.)

A REVIEW OF ONE-WAY AND TWO-WAY EXPERIMENTS TO TEST THE ISOTROPY OF THE SPEED OF LIGHT

Authors: Md. Farid Ahmed, Brendan M. Quine, Stoyan Sargoytchev, and A. D. Stauffer

Abstract

As we approach the 125th anniversary of the Michelson-Morley experiment in 2012, we review experiments that test the isotropy of the speed of light. Previous measurements are categorized into one-way (single-trip) and two-way (round-trip averaged or over closed paths) approaches and the level of experimental verification that these experiments provide is discussed.

The isotropy of the speed of light is one of the postulates of the Special Theory of Relativity (STR) and, consequently, this phenomenon has been subject to considerable experimental scrutiny. Here, we tabulate significant experiments performed since 1881 and attempt to indicate a direction for future investigation.

1. Introduction

In 1905 Albert Einstein introduced the Special Theory of Relativity (STR) – a theoretical framework that proved immediately successful in unifying Maxwell's electrodynamics with classical Mechanics. One of the primary experimental measures of STR was that it provided an explanation for the results of Michelson and Morley's investigation that found no variation in the speed of light with Earth motion.

Under STR, the laws of electrodynamics, as expressed by Maxwell's equations, were held invariant under Lorentz transformations as a consequence of the assumption that the velocity of light is constant in all systems independent of the velocity of the light source. This theory did not only resolve open questions in electrodynamics, it also introduced a revolutionary new notion of space and time as a single entity, space-time.

The main feature of the STR, the space-time symmetry of Local Lorentz Invariance (LLI), has influenced profoundly the development of fields from science-technology to philosophy.

Indeed, our present understanding of all physical theories describing nature are based on Special and General Relativity (GR) – the constancy of the speed of light being necessary for the validity of both relativity theories. LLI is required by GR in the limiting case of negligible gravitation and is today the basis of the standard model of particle physics (relativistic quantum field theory).

Despite the remarkable success of STR and GR several modern theoretical approaches have begun to predict variation on the constant light-speed postulate. String theory which seeks to unify today's standard model with general relativity predicts a violation of the constancy of the speed of light.

Another approach has been described by Zhou and Ma who have proposed a new framework as the Standard Model Supplement (SMS) which brings new terms violating Lorentz invariance in the standard model. Also, Albrecht and Magueijo have proposed the Variable Speed of Light (VSL) theory in order to explain some significant cosmological problems.

However, all theoretical predictions of the violation of the Lorentz Invariance are speculative which lack experimental verification. The widely used experiments to test the STR may be divided into three classical types based on Robertson, and Mansouri and Sexl as: (a) Michelson-Morley (M-M type) which tests the isotropy of the speed of light, (b) Kennedy-Thorndike (K-T type) which tests the velocity dependence of the speed of light, and (c) Ives and Stilwell (I-S type) which tests the relativistic time dilation.

These experiments have been reviewed previously by different authors. Most of these experiments especially M-M type and K-T type only test the two-way speed of light (in a closed path of given length). However, still there are questions about the constancy of the one-way speed of light [24, 25]. Here, we present a comparison and review of the experimental tests which cover isotropy of the velocity of light: one-way and two- way speed of light measurements.

4. Discussion

From today's perspective the constancy of the speed of light influences a variety of areas from science-technology to philosophy. Therefore to accept the idea of the constancy of the speed of light unambiguously, we need experiments sensitive enough to measure the hypothetical violation of the constancy of the speed of light.

The Michelson-Morley experiment is beautiful in its simplicity, but tests only the constancy of the round-trip averaged speed of light. Based on the results of the classic or modern tests of Michelson-Morley experiment as shown in the in **Fig. 6**, <u>we can only establish the special theory of relativity for the round-trip averaged speed of light</u> [underscored by Ahn]. Also we note that Maxwell stated that no apparatus existed capable of measuring effects of the order $\left(\dfrac{v^2}{c^2}\right)$, the square of the ratio of the Earth's speed to that of the light [85]. In order to review isotropy tests of the single-trip (one-way) speed of light, we base our work on Table-1 of the article published by Will in and Table 1 of the article published by Zhou and Ma in, and which are presented in **Fig. 7** in this article. If we compare the one- way experiments of [48] in **Fig. 7** with the two-way experiments in **Fig. 6**, the results

Special Relativity

are about 4 to 6 orders of magnitude smaller in the one-way experiments than those of two-way experiments.

Also the most recent one-way experiment performed by Krisher et al in 1988 in NASA-Jet Propulsion Laboratory Deep Space Network (DSN) presents 2 orders of magnitude smaller values than that of NASA"s previous experiment by Vessot et al in 1976. This is contradictory to our expectation based on STR where we expect lower order of magnitude values with greater improvements. From 1976 to 1988, a twelve year period, science and technology improved and we expect more sensitive and accurate results.

The results of the one-way experiments are increasing in magnitude with time, whereas, the two-way experiments are decreasing in magnitude with greater precision and improvements with time. However, the results from the limits of the one-way experiments of at the GRAAL facility are consistent with STR. But the regularity in the variations of the reported results of the GRAAL measurements reported in different time periods remains unclear and needs further experimental investigations.

At extremely high energy levels the standard model of particle physics and Einstein's general theory of relativity theories coalesce into a single underlying unified theory where the prediction of the violation of the Lorentz invariance at a certain level demands more sensitive experimental tests. We have presented a comparison of experiments in Fig. 8 that shows the one-way speed of light measurement is approximately 2000 times more sensitive than that of round-trip test.

Will [48] showed that experiments which test the isotropy in one-way or two-way (round-trip) have observables that depend on test functions but not on the particular synchronization procedure. He noted that "the synchronization of clocks played no role in the interpretation of experiments provided that one is careful to express the results in terms of physically measurable quantities". Hence the synchronization is largely irrelevant and one-way speed of light is measurable.

Therefore, we would like to propose that not only Michelson-Morley's two-way speed of light measurements be repeated but also other one-way speed of light measurements be performed with greater improvements. Results of the experimental tests spanning at least 24 hours periods in different seasons of the year should be recorded. Any hypothetical diurnal variations that might be observed should follow the figures presented in the section 2.2 in **Fig. 3** and **Fig. 4**.

The whole paper, figures, and references can be found at the following website address:

http://arxiv.org/ftp/arxiv/papers/1011/1011.1318.pdf

Md. Farid Ahmed, a corresponding author of the above paper, can be reached at

E-mail: mdfarid@yorku.ca or

0rdinary mail address: Department of Earth and Space Science and Space Engineering, York University, 4700 Keele Street, Toronto, Ontario, Canada-M3J 1P3.

I would not say that the above paper directly supports my position—the speed (proper speed) of light is constant with respect to *absolute space* and not with respect to human observers. But at least this paper questions the belief of relativists—the speed of light speed is the same for all observers. I do suggest that scientists further their exploration keeping the possibility that the proper speed of light is constant with respect to absolute space (rather than to observers who are arbitrary in motion) in mind. ♦

Special Relativity

22

How to Measure Absolute Motion/ Speed

In the previous chapters I argued that space and time are absolute and that the speed (proper speed) of light is constant with respect to absolute space. Then can we detect or measure the absolute motion/speed of any object with respect to absolute space? I think we can.

We make a light source that emits intermittent light pulses, and we put such an apparatus on the nose (or tail) of a spacecraft (or any other object) in space. If the spacecraft is stationary with respect to absolute space, the light pulses from the "intermittent-light-pulse emitter" would propagate in all direction at the same seed (c), and so the pulses will make a series of *concentric circles* around the light source as shown in diagram (a) in **Fig. 22-1**.

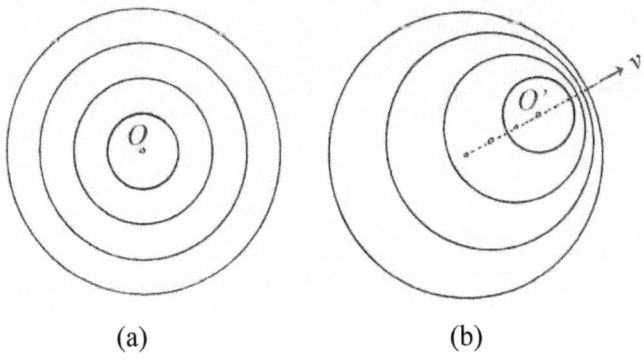

(a)   (b)

Fig. 22-1 Source *O* is stationary in (a). Source O is in motion in (b).

However, if the fronts of the light pulses make *eccentric circles* as shown in diagram (b) of **Fig. 22-1**, it means that the source is in motion at a certain velocity (*speed* plus *direction*) with respect to absolute space. In this way we might find the absolute speed of objects. This is a theoretical method and not a practical method because we cannot gather the information of the propagation of light pulses.

Quite recently (August 2013) I read a paper which says that scientists have developed the technology of measuring "one-way light speed" (see **Chapter 21**). [1] So we can detect the direction and he speed of absolute motion of the spacecraft (or any object) by gathering and analyzing the information of the one-way light speeds in several directions from the light source attached to the spacecraft.

The above method is based on the idea that the *proper speed of light* is invariable with respect to space (absolute space). But one thing I am not sure of is that how light behaves in transparent mediums such as air, glass, water, or *Vacuum Island* trapped in a laboratory. Will the transparent mediums or vacuum islands impede or muffle the effect of the absoluteness of space in varying degrees? To what degree could the presence of air or the turbulent air/atmosphere of the earth affect the speed of light in the Michelson-Morley interferometer? I think scientists can find the answers to these questions. Michelson-Morley experiment was not complete null effect (see **Chapter 27**). ♦

Special Relativity

23

Relativistic Addition of Speeds

Relativists think that the relativistic addition of speed (relative speed or proper speed?) proves the invariance of light speed. In this chapter, I will prove that the theory of invariance of light speed is wrong by showing the problem with the relativistic addition of speeds. This problem has to do with not distinguishing relative speed from proper speed. Let us do some simple exercises:

Exercise 23-1

In **Fig. 23-1**, photon P_1 and photon P_2 have started from common source (or observer) O at the same time and have moved in the opposite directions of each other for one second. Assume that source (or observer) O is stationary in free space.

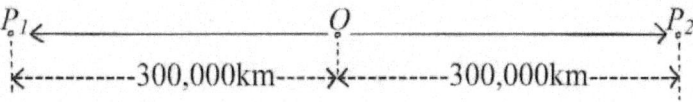

Fig. 23-1

Questions:

Find the following from the Newton-Galilean point of view and from the Einsteinian point of view.

1. The distance between the two photons ($P_1\ P_2$).
2. The relative speed between the two photons.

Newton-Galilean solution:

Harvard Physicists Confuse Relative Speed with Proper Speed

In **Fig. 23-1**, $OP_1 = OP_2 = 300{,}000$km.
Therefore, $P_1 P_2 = 2 \times 300{,}000$km.

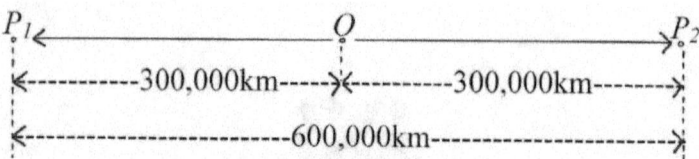

Fig. 23-2 Newton-Galilean point of view.

The relative speed between P_1 and P_2 is $2 \times 300{,}000$km/s $= 2c$. (Or simply $c + c = 2c$.)

Note that 2c is the *relative speed* between the two photons. Yet the proper speed of each photon is, as always is, c ($= 300{,}000$ km/s).

Einsteinian solution:

(1) According to the relativistic addition of speeds, the relative speed between two photons P_1 and P_2 is $\dfrac{c+c}{1+\dfrac{c\times c}{c^2}} = c$.

Therefore, distance $P_1P_2 = c \times 1\text{sec} = 300{,}000$ km.

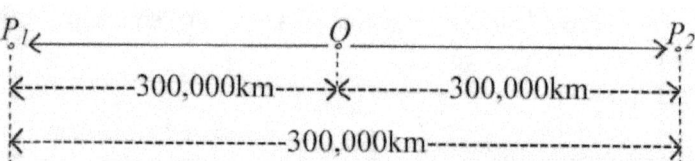

Fig. 23-3 Einsteinian point of view-1

The result $P_1P_2 = c \times 1\text{sec} = 300{,}000$km (see **Fig. 23-3**) is not of fully relativistic view because this solution does not reflect the distance contraction effect.

Let us apply the *distance contraction* effect to this case as relativists do in the twin paradox (see **Chapter 29**).

$P_1P_2 = 300{,}000$km (or $600{,}000$ km) $\times \sqrt{1 - c^2/c^2} = 0$.
(This is when viewed from observer O.)
Likewise,

Special Relativity

$OP_1 = 300{,}000\text{km} \times \sqrt{1 - c^2/c^2} = 0$, and

$OP_2 = 300{,}000\text{km} \times \sqrt{1 - c^2/c^2} = 0$.

(These results are when viewed from observer O with the distance contraction effect being applied. See **Fig. 23-4**.)

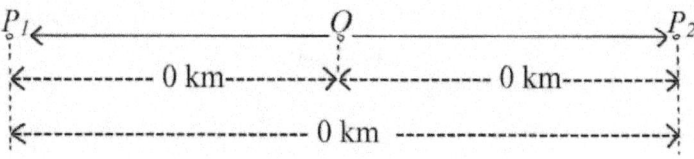

Fig. 23-4 Einsteinian point of view-2

However, if we are to apply the *distance elongation effect to this case* as do relativists to the *muon paradox* (see **Chapter 30**),

$P_1P_2 = 600{,}000\text{km} \times \dfrac{1}{\sqrt{1 - c^2/c^2}} = \infty \text{ km}.$

Likewise,

$OP_1 = 300{,}000\text{km} \times \dfrac{1}{\sqrt{1 - c^2/c^2}} = \infty \text{ km}$, and

$OP_2 = 300{,}000\text{km} \times \dfrac{1}{\sqrt{1 - c^2/c^2}} = \infty \text{ km}.$

(These results are when observed from observer O with the distance elongation effect being applied. See **Fig. 23-5**.)

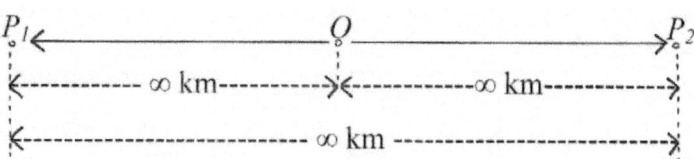

Fig. 23-5 Einsteinian point of view-3

So far, we can see that Newton-Galilean solution makes sense while Einsteinian solution not only does not but also is inconsistent.

Exercise 23-2

Harvard Physicists Confuse Relative Speed with Proper Speed

In **Fig. 23-6**, two photons A and B have started from sources O_A and O_B respectively at the same time and have moved in the same direction for one second in free space. Distance $O_A O_B$ is 5cm.

O_A _____ O_B _____ A _____ B

|<--- 5cm --->|

Fig. 23-6

Questions:

Find the following from the Newton-Galilean point of view and then from the Einsteinian point of view.

1. Distances $O_A A$, $O_B B$, and AB.
2. The relative speed between the two photons (v_{AB}).

For convenience's sake, substitute "$c \times 1\text{sec}$" with k where c is the speed of light (300,000 km/sec) and k = 300,000 km.

Newton-Galilean solution:

$O_A A$ = 300,000km/sec × 1sec = 300,000 km = k.
OBB = 300,000km/sec × 1sec = 300,000 km = k.
AB = 5cm.
The relative speed between A and B (v_{AB}) = $c - c$ = 0.

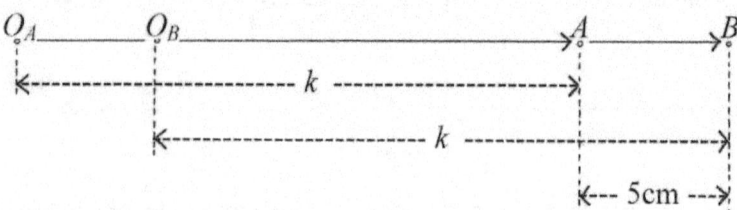

Fig. 23--7 Newton-Galilean point of view.

Einsteinian solution:

$O_A A$ = 300,000km/sec × 1sec = 300,000 km = k.
$O_B B$ = 300,000km/sec × 1sec = 300,000 km = k.
AB =?

Special Relativity

If we are to find distance AB in relativistic view point, we need to find the relative speed between A and B; According to the relativistic addition of speeds, the relative speed between the two photons (v_{AB}) is

$$\frac{c-c}{1+\frac{c \times (-c)}{c^2}} = \frac{c-c}{c^2-c^2} = \frac{0}{0} = ?$$
$$\frac{}{c^2}$$

This is not what Einstein means.

We need some manipulation to find the relativistic solution.
Let us replace the speed of photon B with "ϕ" where
$\phi = \lim\limits_{n \to \infty} c\{1-(0.1)^n\}$. Note that $\lim\limits_{n \to \infty} c\{1-(0.1)^n\}$ is actually c. Then,

$$v_{AB} = \frac{c-\phi}{1+\frac{c \times (-\phi)}{c^2}} = \frac{c-\phi}{c-\phi} = c. \text{ (This is what relativity means.)}$$

According to the above result,
$AB = 5\text{cm} + c \times 1\sec = 5\text{cm} + k.$

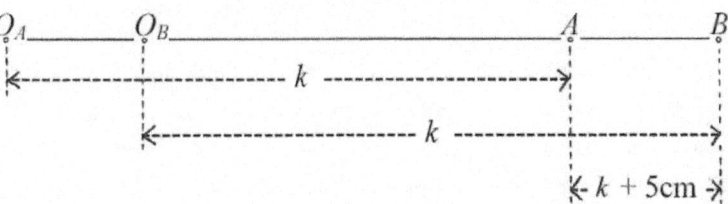

Fig. 23-8 Einsteinian point of view

The above result (see **Fig. 23-8**) is not fully relativistic because this does not reflect the distance contraction or distance elongation effect. But let us omit the application of these effects. We can see that the relativistic addition of speed is contradictory.

Let us do one more exercise to see the dramatic fallacy of the relativistic speed addition:

Exercise 23-3

In **Fig. 23-9**, two photons A and B and *God* have started from common origin O at the same time, and are moving in the same straight line in free space. Photon B and God move in the same direction, and photon A moves in the opposite direction of photon B and God. Assume that origin O is stationary in free space and that God moves at 1000 times the speed of light ($=1000c$).

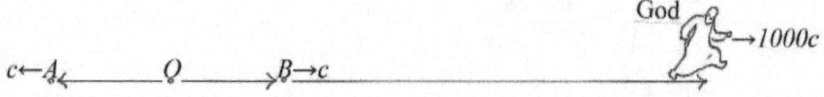

Fig. 23-9 God moves at the speed of $1000c$.

According to Einstein, God cannot move faster than light speed. But if that were true, God cannot govern the whole universe whose size seems infinite. We just assume that God moves at the speed $1000c$ for our thought experiment. (The assumption that God's speed is $1000\ c$ is still a derogative idea for believers. God's speed should be infinite (∞) times the speed of light.) Assume again that photons A, B, and God have moved for one second after they started from common origin O.

Questions:

Find the following from the Newton-Galilean point of view and then from the Einsteinian point of view:

(1) The relative speeds between A and B, A and God, and B and God.
(2) The distances between A and B, A and God, and B and God.
For convenience's sake, substitute "$c \times 1\mathrm{sec}$" with k where c is the speed of light ($=300{,}000$ km/sec) and $k = 300{,}000$ km.

Newton-Galilean point of view:

Fig. 23-10 Newton-Galilean point of view.

Fig. 23-10 shows the followings:
   The relative speed between A and B $= c + c = 2c$.
   The relative speed between A and God $= 1000c + c = 1001c$.

168

Special Relativity

The relative speed between B and God = $1000c - c = 999c$.

According to the above results,

Distance $AB = 2c \times 1\text{sec} = 2k$.
Distance A-God $= 1001c \times 1\text{sec} = 1001k$.
Distance B-God $= 999c \times 1\text{sec} = 999k$.

(II) Einsteinian point of view:

The relative speed between A and B =

$$c + c = \frac{c+c}{1+\frac{c \times c}{c^2}} = c$$

The relative speed between A and God =

$$c + 1000c = \frac{c+1000c}{1+\frac{c \times 1000c}{c^2}} = \frac{1001c}{1001} = c.$$

The relative speed between B and God =

$$1000c - c = \frac{c - 1000c}{1+\frac{c \times (-1000c)}{c^2}} = \frac{-999c}{-999} = c.$$

[Note that the sign of God's speed is negative (–) because the direction of God is the same as the direction of photon B.]

According to the above results,

Distance $AB = c \times 1\text{sec} = k$.
Distance A-God $= c \times 1\text{sec} = k$.
Distance B-God $= c \times 1\text{sec} = k$.

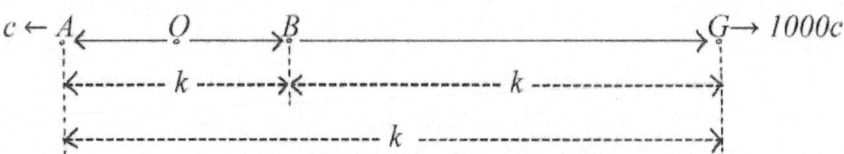

Fig. 23-11 Einsteinian point of view

What happened to God and photon K? How can the relative speed of God, whose proper speed is $1000c$, with respect to the photon, whose proper speed is c, only c? The answer is simple as follows:

In general, the result of the following expression is always c **no matter** what value the "any" in the following expression may be:

$$\frac{c \pm any}{1 + \frac{c \times (\pm any)}{c^2}} = \frac{c \pm any}{\frac{c(c \pm any)}{c^2}} = c.$$

It is not surprising that the relativistic formula of speed addition makes the speed of light always the same (c) because the **Lorentz transformation**, from which the formula was derived, was made for the sole purpose of making the speed of light (or electromagnetic waves) always c. So the act of proving the constancy of light speed with the formula of speed addition is but a *circular reasoning*.

Relativists believe that the constancy of light speed was proved centuries ago through actual experiments or observations by Roemer (1644-1710), Bradley (1693-1762), Fizeau (1819-1896), etc. But these scientists did not discern the proper speed from the relative speed of light. Maxwell (1831-1879) predicted the theoretical speed of light (c) and the constancy of the speed of electromagnetic waves. But he believed that the speed of light is constant with respect to absolute space allegedly filled with ether. So Maxwell's light speed was the proper speed and not the relative speed. Michelson (1852-1931) was famous for his accurate measurements of "two-way light speed." But what he measured was the "average proper speed" of light in 1-D situations. Michelson did not believe in the theory of relativity (see **Chapter 27**). [1]

The ability to distinguish proper speed from relative speed is critically important in physics. Einstein invented the theory of relativity on the lack of this ability. The upper limit of speeds came from the LT. According to the Lorentz factor $\sqrt{1 - v^2/c^2}$, no object can exceed the speed of light (c). When it comes to proper speed, perhaps light speed is the upper limit of speeds. But when it comes to relative speed, even a human or a snail can move at the speed of light or faster-than-light-speed. For example, if a snail and a photon move in the opposite direction of each other on the same straight line with proper speeds of v and c respectively in space, the relative speed of the snail with respect to the photon is $v + c$, and *vice versa*. How do we measure the proper speeds of objects with respect to space (absolute space)? I suggested a method of

Special Relativity

finding the *absolute speed*s of objects in **Chapter 22**.

How Did Einstein come to believe the invariance of light speed?

Decades before Einstein formulated his theory of relativity, Maxwell (1831-1879) predicted, based on his equations, that electromagnetic waves should exist that travel at the speed of light. [Maxwell also surmised that light consisted of electromagnetic waves. Maxwell's idea was proved correct after he died. [1]

Maxwell thought, as many of his contemporaries did, that space was filled with ether and that space was absolute. Therefore, the speed of light Maxwell mentioned was the proper speed of light in absolute space. Many physicists in Maxwell's time, including Maxwell himself, did not distinguish proper speed from relative speed. They just thought that all motion/speed is relative and that the speed of light is the same for all observers. They were not aware of the difference between proper speed and relative speed.

After I saw many problems with relativity, I came to think that what is constant is the *proper speed* of light with respect to absolute space and that the relative speed of light with respect to specific observers may vary depending on the motion or/and position of observers. Sagnac experiment (1913) and Einstein's train thought experiment concerning the relativity of simultaneity show that the relative speed (not proper speed) of light with respect the observer/interferometer does vary (see **Chapter 15**).

Before Einstein invented special relativity, he had agonized a lot over the idea of why the speed of light is the same for all observers. [2] Any one of commonsense should agonize over such an unphysical assertion that the speed of light is the same for all observers. Modern physicists do not agonize over such an idea or postulation. Modern physicists have stopped inquisitive questioning. Since Einstein did not distinguish proper speed from relative speed, his agony did not help him solve the basic problem. After severe agony over the constancy of light speed Einstein just concluded that time and space contract or expand. Einstein got the wrong answer: he just followed the contemporary ideas that light speed is constant and that time and space are relative entities.

Lorentz had invented a set of equations called the Lorentz transformation (LT) in order to explain (rationalize) the invariance of

> light speed. Einstein borrowed the LT, and he thought that space and time expand or contract to make the speed of light always the same. Therefore, Einstein's theory was nothing new.

Most literature that deals with the history of relativity mentions that many scientists had found, through many observations or experiments, that the speed of light is the same for all observers. But this conclusion is based on the inability to discern relative speed from proper speed. Modern physicists cannot distinguish proper speed from relative speed due to the absolute dogma of Einstein. ♦

Special Relativity

24

The Lorentz Transformation (LT)

Lorentz (1853-1928) found a set of equations called Lorentz transformation (LT) in 1895. The purpose of the LT was to explain the *invariance of light speed*.

Later, in 1905, Einstein formulated special relativity borrowing the math of the Lorentz transformation. Einstein believed that all motion is relative and that the speed of light is the same for all observers. Here we can see that Einstein's two postulates are contradictory with each:

It should be pointed out that the LT does not deal with the relative speed; it is to translate the proper speed of an object or photon in an inertial system in to the proper speed of the same object or photon in another inertial system which is in uniform motion relative to the first.

Arthur Beiser explains the Lorentz transformation in his textbook entitled *Concepts of Modern Physics* as follow: [1] (In the following citation, the equation numbers are from Beisr's textbook. But the figure numbers are reset to fit this book. The words in brackets are mine.)

> Suppose we are in an inertial frame of reference S and find the coordinates of some event that occurs at the time t are x, y, z. An observer located in a different inertial frame S' which is moving with respect to S at the constant velocity v will find that the same event occurs at the time t' and has the coordinates x', y', z'. (In order to simply our work, we shall assume that v is in the $+x$ direction, as in **Fig. 24-1**.) How are the measurements x, y, z, t related to x', y', z', t'?
> **Galilean Transformation**

Before special relativity, transforming measurements from one inertial system to another seems obvious. If clocks in both systems are started when the origins of S to S' coincide, measurements in the x direction made in S will greater than those made in S' by the amount vt, which is the distance S' has moved in the x direction. That is,

$$x' = x - vt \qquad (1.26)$$

There is no relative motion in the y and z directions, and so

$$y' = y \qquad (1.27)$$
$$z' = z \qquad (1.28)$$

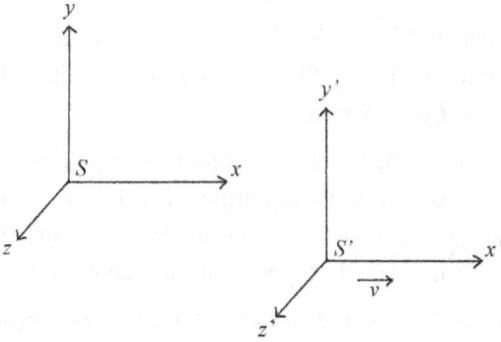

Fig. 24-1 Frame S' moves in the $+x$ direction with the speed v relative to frame S. The Lorentz transformation must be used to convert measurements made in one of these frames to their equivalents in other.

In the absence of any indication to the contrary in our everyday experience, we further assume that

$$t' = t \qquad (1.29)$$

The set of **Eqs. (1.26)** to **(1.29)** is known as the **Galilean transformation**.

To convert velocity components measured in the S frame to their equivalents in the S' frame according to the Galilean transformation, we simply differentiate x', y', and z' with respect to time t:

$$v'_x = \frac{d'_x}{dt'} = v_x - v \qquad (1.30)$$

Special Relativity

$$v'_y = \frac{d'_y}{dt'} = v_y \qquad (1.31)$$

$$v'_z = \frac{d'_z}{dt'} = v_z \qquad (1.32)$$

Although the Galilean transformation and the corresponding velocity transformation seem straightforward enough, they violate both of the postulates of special relativity. The first postulate calls for the same equations of physics in both the S and S' inertial frames, but the equations of electricity and magnetism become very different when the Galilean transformation is used to convert quantities measured in one frame into their equivalents in the other. The second postulate calls for the same value of the speed of light c whether determined in S or S'. If we measure the speed of light in the x direction in the S system to be c, however, in the S' system it will be $c' = c - v$

according to **Eq. (1.30)**. Clearly a different transformation is required if the postulates of special relativity are to be satisfied. We would expect both time dilation and length contraction to follow naturally from this new transformation.

Lorentz Transformation

A reasonable guess about the nature of the correct relationship between x and x' is

$$x' = k(x - vt) \qquad (1.33)$$

Here k is a factor that does not depend upon either x or t but may be a function of v. The choice of **Eq. (1.33)** follows from several considerations:

1. It is linear in x and x', so that a single event in frame S corresponds to a single event in frame S', as it must.

2. It is simple, and a simple solution to a problem should always be explored first.

3. It has the possibility of reducing to **Eq. (1.26)**, which we know to be correct in ordinary mechanics.

Lorentz found that the factor k is

$$k = \frac{1}{\sqrt{1 - v^2/c^2}} \qquad (1.40)$$

[The procedure of finding factor k is omitted here for brevity. See

Beiser's textbook.]

The complete transformation of measurements of an event made in S to corresponding measurements made in S' are as follows:

$$x' = \frac{x - vt}{\sqrt{1 - v^2/c^2}} \qquad (1.41)$$

$$y' = y \qquad (1.42)$$

$$z' = z \qquad (1.43)$$

$$t' = \frac{t - \frac{vx}{c^2}}{\sqrt{1 - v^2/c^2}} \qquad (1.44)$$

In Fig. 24-1, the origin of motion in S frame coincides with that of S' frame. This means that Lorentz considered the speed of an object from its origin of motion. The origin of motion is assumed to be stationary with respect to the given inertial coordinate system (S or S'). This mean Lorentz dealt with proper speed, which is independent, and not relative speed, which is observer-dependent; observers are obsolete in the LT. Lorentz (and Einstein) did not distinguish the concept of proper speed from relative speed.

To sum up, the LT is a mathematical means (manipulation) to translate the proper speed of an object (or photon) in one inertial frame of reference into the proper speed of the same object in another frame of reference moving relative to the first **under the presumption that** the speed (*proper speed!*) of light is the same in any frame of reference.

The proper speed is one-dimensional in concept because the proper speed of an object is measured with respect to an observer (or speed detector) who sits on the line of motion of the object (see **Chapter 1**). So the LT is a one-dimensional consideration (no matter in what direction an object or photon moves). The sole purpose of the LT is to rationalize the isotropy of light speed. The proper speed of light is the constant (c) with respect to **absolute space** and not with respect to any inertial frames/observers (see **Chapter 21**).

The formula of addition of speed, which is the derivation from the LT, really shows that $c \pm v = c$. This is because the LT was formulated based on the postulate that $c \pm v = c$. So relativists and physics students

Special Relativity

think that the speed of light is truly the same for all observers. But this is only circular reasoning.

Relativists believe that the Michelson-Morley experiment proved that the speed of light is the same for all observers and that there is no ether. But Michelson did not agree with the relativistic interpretation of his experiment. Arthur Beiser states about the Lorentz transformation and the Michelson-Morley experiment as follows: [2]

> The set of equations that enables electromagnetic quantities in one frame of reference to be transformed into their values in another frame of reference moving relative to the first were found by Lorentz in 1895, although their full significance was not realized until Einstein's theory of special relativity 10 y afterward. Lorentz (and independently, the Irish physicists G.F. Fitzgerald) suggested that the negative result of the Michelson-Morley experiment could be understood if lengths in the direction of motion relative to an observer were contracted. Subsequent experiments showed that although such contractions do occur, they are not the real reason for the Michelson-Morley result, which is that there is no "either" to serve as a universal frame of reference.

Lorentz believed in ether. He believed that if an object moves in ether the length of the object is shortened (contracted) in the direction of motion by the factor of $\sqrt{1-v^2/c^2}$ where v is the speed of the object and c is the speed of light. We should note that v and c are the absolute speed with respect to ether, which contemporary physicists believed as the absolute frame of reference. That is, v and c have nothing to do with observers.

The LT was formulated based on the postulate that the speed of light is the same for all observers. That is, $c \pm v = c$. But the Lorentz factor ($\sqrt{1-v^2/c^2}$ or $\frac{1}{\sqrt{1-v^2/c^2}}$), which is derived from the LT, is self-contradictory because $(c-v) = (c+v) = c$. Then,

$$\sqrt{1-v^2/c^2} = \sqrt{\frac{(c+v)(c-v)}{c^2}} = \sqrt{\frac{c \times c}{c^2}} = 1, \text{ and}$$

$$\frac{1}{\sqrt{1-v^2/c^2}} = 1.$$

So the LT, Lorentz factor, the isotropy of light, time dilation, length contraction, mass increase, etc. are wrong.

At least Lorentz could explain (rationalize) the length/distance contraction effect (in the direction of motion) by assuming the resistance (pressure) of ether. Einstein borrowed the LT and the idea of the length contraction, time dilation effect. But Einstein discarded the ether. If there is no ether, as Einstein believed at the time he formulated special relativity, there is no reason for a moving object to be contracted in the direction of its motion. Einstein found length contraction, time dilation, and mass increase *theoretically* or *mathematically* not *experimentally*.

Another fundamental problem with the LT is that Lorentz and Einstein thought that the pattern of uniform motion of ordinary objects such as a train, car, spacecraft, etc. is basically the same as the pattern of motion (propagation) of photon or electromagnetic waves. The pattern of motion of photons is not the same sort as that of ordinary objects. The motion of electromagnetic waves and that of ordinary objects cannot be unified by math (see **Chapter 17**). ♦

Special Relativity

25

The Light-year and Relativity

The unit *light-year* is the distance light traverses for one year in rectilinear direction in space. This distance is about 9.46 trillion kilometers or 5.88 trillion miles. We should note that this unit is based on Newton-Galilean physics and not modern physics (relativity). If we are to consider time dilation or distance/length contraction effect, light-year should be redefined as follows:

According to distance contraction effect of relativity (see twin paradox in **Chapter 29**), the distance a photon travels for one year is reduced to zero (0).

$$1 \text{ light-year} \times \sqrt{c - c^2/c^2} = 0 \text{ cm.}$$

(This is when viewed from an observer who is aligned with the line of motion of the photon. For those observers who are off the line of motion of the photon, the distance contraction rare will different the above result.)

According to time dilation effect (see twin paradox in **Chapter 29**), one year is dilated to zero (0).

$$1 \text{ year} \times \sqrt{c - c^2/c^2} = 0 \text{ second}$$

(This is when viewed from an observer who is aligned with the line of motion of the photon. The word that the time flow rate is dilated to zero means that the photon never ages. In other word, one second becomes eternity. This is the view point of observers who are aligned with the line of motion of the photon.)

The above stories should be real for relativists who believe in distance contraction and time dilation. (I would omit the mass increase effect here.)

Relativists may say that relativistic effects are not applied to light (photon) itself. But that kind of argument is groundless. The law of physics should be applied to all.

Assuming that the above stories are all real, let us do an exercise to see how relativity is unbelievable.

Exercise 25-1

Suppose that three stars *A*, *B*, and *C* are one light-year, one million light-years, and one hundred million light-years of distances from the earth respectively (see **Fig. 25-1**). Suppose again that Einstein on the earth sends light signals to each of these three stars at the same time.

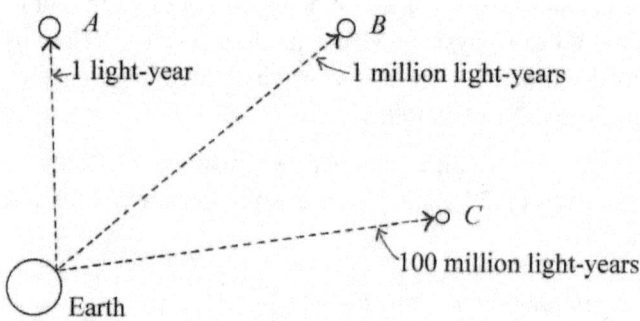

Fig. 25-1 Einstein sends light signal to each of the three stars at the same time.

Questions:

How long will the light signals appear to take to get to each of these stars?

Find the answer from the Galilean point of view and then from the relativistic point of view. (Assume that the positions of the four celestial bodies (the earth and the three stars) are stationary in free space.)

Galilean Solution:

It will take the light signal one light year to get to star *A*.

It will take the light signal one million years to get to star *B*.

Special Relativity

It will take the light signal 100 million years to get to star C.

Relativistic Solution:

To get to star A, it will take the light signal 1 year $\times \sqrt{1-c^2/c^2} = 0$ year (zero second).

To get to star B, it will take the light signal one million years $\times \sqrt{1-c^2/c^2} = 0$ year (zero second).

To get to star C, it will take the light signal 100 million light-years $\times \sqrt{1-c^2/c^2} = 0$ year (zero second).

(The above result means that the distance between the earth and each of the three stars is reduced to zero km or zero cm when viewed from Einstein on the earth.)

The Lorentz factor $\sqrt{1-c^2/c^2}$ is what relativists apply to the distance contraction effect in the twin paradox (see **Chapter 29**). However, according to the distance elongation effect which relativists apply to the muon paradox (see **Chapter 30**), the distance between the earth and each of the three stars would be elongated at the rate of $\dfrac{1}{\sqrt{1-v^2/c^2}}$ as follows:

The distance between the earth and star A = one light year $\times \dfrac{1}{\sqrt{1-c^2/c^2}} = \infty$ light year.

The distance between the earth and star B = one million light years $\times \dfrac{1}{\sqrt{1-c^2/c^2}} = \infty$ light years.

The distance between the earth and star C = 100 million light years $\times \dfrac{1}{\sqrt{1-c^2/c^2}} = \infty$ light years.

And the time the light signals take to get to each of the three stars would be as follows:

To get to star A, it will take the light signal 1 year $\times \dfrac{1}{\sqrt{1-c^2/c^2}} = \infty$ year.

To get to star B, it will take the light signal 1 million year $\times \dfrac{1}{\sqrt{1-c^2/c^2}} = \infty$ years.

To get to star C, it will take the light signal 100 million light year $\times \dfrac{1}{\sqrt{1-c^2/c^2}} = \infty$ years.

These stories show that relativity is inconsistent and not true.

The concept of light-year is of classical physics. This concept is not compatible with relativity. It is not proper for modern physicists (relativists) to use the unit *light-year* in modern physics because it is a Newton-Galilean unit. ♦

Special Relativity

26

Does the Universe Expand at Faster-than-light Speed?

According to the Big Bang theory, the universe keeps expanding from the beginning of the universe (the Big Bang) and the frontier or outermost part of the universe is expanding at the faster-than-light speed. In modern astronomers, the term Hubble's horizon is defined as the boundary of the universe from which the expansion speed of space is the same as speed of light.

Aside from the controversy over the Big Bang theory itself, here are two major questions concerning the speed of expansion of speed. The first question is this: Relative to what does the Hubble's horizon or outermost part of the universe expand at the light speed or faster-than-light speed? The second question is how the universe can expand at the faster-than-light speed. Is it not contradictory that the speed of expansion of space (universe) transcends the speed of light?

The answer to the first question is possible if space is absolute and heavenly bodies at the outermost part are in motion at whatever speed with respect to the absolute space. I mean if relativists' space-time is relative, how can relativists say that the space itself is expanding at the speed of light or faster? Relativists say that all motion/speed is relative. I do want to hear from relativists, especially Harvard physicists, about *relative to WHAT* the universe expands at the faster-than-light speed.

The meaning of the word "move at the light speed or faster-than-light speed ..." makes sense if space itself is absolute and heavenly bodies at the outermost part of the universe are in motion at so and so speed (*actual speed*, not *relative speed*, see **Chapter 1**) with respect to absolute space. I proved that space and time are absolute in **Chapters 18-19**. But I do not mean that universe is truly expanding at the faster-than light speed or that the Big Bang theory is true. I think that space is absolute and limitless and that regional expansion of galaxies is possible but the idea that the frontier part of the entire universe is expanding at the exponential speed is impossible because it is against the law of conservation of energy.

The second question is how relativists, who believe that light speed (c) is the upper limit of speed, recognize the faster-than-light speed in astronomy.

According to modern astronomy, the frontier part of the universe is moving away at an exponential speed. That means the heavenly bodies at the forefront of the frontier of the universe move at the speed much faster than light speed, perhaps thousands or millions times the speed of light or faster.

Relativistic astrophysicists explain the faster-than-light speed of expansion of universe in a quite strange way. Wikipedia, the internet encyclopedia, states about the faster-than-light speed as follows: [1]

Faster-than-light observations and experiments

There are situations in which it may seem that matter, energy, or information travels at speeds greater than c, but they do not. For example, as is discussed in the propagation of light in a medium section below, many wave velocities can exceed c. For example, the phase velocity of X-rays through most glasses can routinely exceed c, but such waves do not convey any information.

If a laser beam is swept quickly across a distant object, the spot of light can move faster than c, although the initial movement of the spot is delayed because of the time it takes light to get to the distant object at the speed of c. However, the only physical entities that are moving are the laser and its emitted light, which travels at the speed of c from the laser to the various positions of the spot. Similarly, a shadow projected onto a distant object can be made to move faster than c, after a delay in time. In neither case does any matter, energy, or information travel faster than light.

The rate of change in distance between two objects in a frame of reference with respect to which both are moving (their closing speed) may have a value in excess of c. However, this does not represent the speed of any single object as measured in a single

Special Relativity

inertial frame.

Certain quantum effects appear to be transmitted instantaneously and therefore faster than c, as in the EPR paradox. An example involves the quantum states of two particles that can be entangled. Until either of the particles is observed, they exist in a superposition of two quantum states. If the particles are separated and one particle's quantum state is observed, the other particle's quantum state is determined instantaneously (i.e., faster than light could travel from one particle to the other). However, it is impossible to control which quantum state the first particle will take on when it is observed, so information cannot be transmitted in this manner.

Another quantum effect that predicts the occurrence of faster-than-light speeds is called the Hartman effect; under certain conditions the time needed for a virtual particle to tunnel through a barrier is constant, regardless of the thickness of the barrier. This could result in a virtual particle crossing a large gap faster-than-light. However, no information can be sent using this effect.

So-called superluminal motion is seen in certain astronomical objects, such as the relativistic jets of radio galaxies and quasars. However, these jets are not moving at speeds in excess of the speed of light: the apparent superluminal motion is a projection effect caused by objects moving near the speed of light and approaching Earth at a small angle to the line of sight: since the light which was emitted when the jet was farther away took longer to reach the Earth, the time between two successive observations corresponds to a longer time between the instants at which the light rays were emitted.

In models of the expanding universe, the farther galaxies are from each other, the faster they drift apart. This receding is not due to motion through space, but rather to the expansion of space itself. For example, galaxies far away from Earth appear to be moving away from the Earth with a speed proportional to their distances. Beyond a boundary called the Hubble sphere, the rate at which their distance from Earth increases becomes greater than the speed of light.

In September 2011, physicists working on the OPERA experiment published results that suggested beams of neutrinos had travelled from CERN (in Geneva, Switzerland) to LNGS (at the Gran Sasso, Italy) faster than the speed of light. These findings, sometimes referred to as the faster-than-light neutrino anomaly, were subsequently determined—subject to further confirmation—to be the result of a measurement error.

Reference numbers and references of the above citation are omitted. These are found at the following address:

https://en.wikipedia.org/wiki/Speed_of_light.

The author of the above citation confuses *proper speed* or *actual speed* (which is observer-free) with *relative speed*. So the above statement is a jargon which reveals that the author does not know what relative speed is.

When it comes to **proper speed**, the speed of light c seems to be the upper limit of speeds. But when it comes to **relative speed**, the speed of light can be from *zero* (0) *to* 2c (two times the speed of light) see the illustration in **Chapter 23**).

The theory of limitless-speed of expansion of the universe (the Big Bang theory) is groundless. Where does the endless power (energy) that allegedly accelerates the speed of heavenly bodies come from? Some relativists explain that the frontier of the universe is not expanding actively but is simply sucked up, as dust is sucked up by a vacuum cleaner, toward the outward direction of space. This argument recognizes, inadvertently, the vacuous space outside the frontier of the universe. As far as we know, most portion of space in our universe is vacuum or quasi-vacuum. Then why are heavenly bodies sucked up only in the outward direction of the universe? Therefore the vacuum-suction theory is wrong.

Relativists take the red-shift of starlight as the evidence of the expansion of universe. But there are many other explanations of the red-shift. The universe might expand locally. But it is not likely that the entire universe keeps expanding forever at an exponential rate. ♦

Special Relativity

27

The Michelson-Morley Experiment

Relativists explain that the Michelson-Morley experiment (MMX, 1887) proved the non-existence of the ether, the relativeness of space, and the invariance of light speed. It is true that the MMX proved the non-existence of the ether. But it is wrong to say that the MMX proved the invariance of light speed and the relativeness of space.

Let us examine what the MMX truly proved or disproved and how relativists made some mistakes in interpreting the experiment. Arthur Beiser explains the MMX in his textbook as follows: [1] (The figure number is reset for this book.)

> Michelson's most significant achievement, carried out in 1887 in collaboration with Edward Morley, was an experiment to measure the motion of earth through the "ether," a hypothetical medium pervading the universe in which light waves were supposed to occur. The motion of the ether was a hangover from the days before light waves were recognized as electro-magnetic, but nobody at the time seemed willing to discard the idea that light propagates relative to some sort of universal frame of reference.
>
> To look for the earth's motion through the ether, Michelson and Morley used a pair of light beams formed by a half-silvered mirror, as in **Fig. 27-1**. One light beam is directed to a mirror along a path perpendicular to the ether current, and the other goes to a mirror along a path parallel to the ether current. Both beams end up at the same viewing screen. The clear glass plate ensures that both beams pass through the same thickness of air and glass. If the transit times of the two beams are the same, they will arrive at the screen in phase and will interfere constructively. An ether

current due to the earth's motion parallel to one of the beams, however, would cause the beams to have different transit times and the result would be destructive interference at the screen. This is the essence of the experiment.

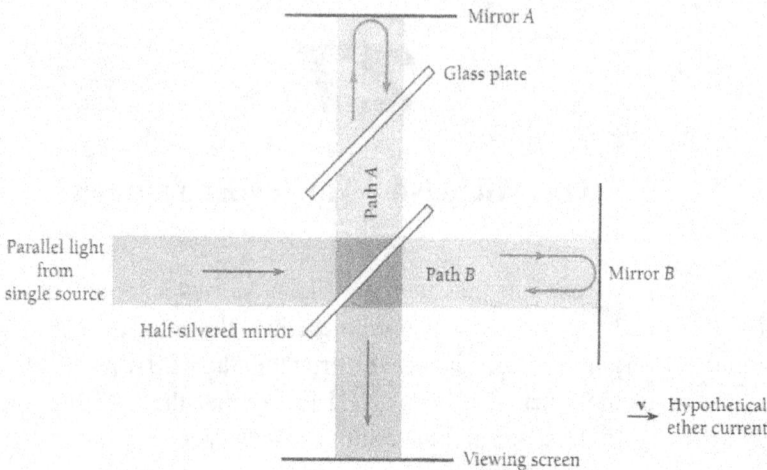

Fig. 27-1 Michelson-Morley experiment

Although the experiment was sensitive enough to detect the expected ether drift, to everyone's surprise none was found. The negative result had two consequences. First, it showed that the ether does not exist and so there is no such thing as "absolute motion" relative to the ether: all motion is relative to a specified frame of reference, not to a universal one. Second, the result showed that the speed of light is the same for all observers which is not true of waves that need a material medium in which to occur (such as sound and water waves).

The Michelson-Morley experiment set the stage for Einstein's 1905 special theory of relativity, a theory that Michelson himself was reluctant to accept. Indeed, not long before the flowering of relativity and quantum theory revolutionized physics, Michelson announced that "physical discoveries in the future are a matter of the sixth decimal place." This was a common opinion of the time. Michelson received a Nobel Prize in 1907, the first American to do so.

Fig. 27-1 is a simplified diagram explaining the concept of the

Special Relativity

MME. The actual MME setup was much more complicated. Michelson believed in the material ether and his purpose was to confirm the ether wind or the entrainment effect of ether. And he basically did not believe in relativity.

Though relativists emphasize the null effect, it was not complete so; some degrees of irregular interferences were detected though smaller than expected. [2]

In any event, Michelson's experiments were interpreted by relativists, who got the absolute advantage of power in numbers, as the evidence that there was/is no ether and that the speed of light is the same for all observers. Yet Michelson did not accept relativists' interpretation.

I would point out several aspects concerning the MMX experiments.

The MMX experiments were performed in the air-filled laboratory on the ground. My position is that space retains the absoluteness (concerning the propagation speed of light) without the ether, and that the absoluteness of space might have been affected by the presence of air or the turbulence (wind) of air.

We should note that the presence of air or the turbulence of atmosphere (wind) might have affected the MMX though the effects were smaller than expected. I think that the speed of light is constant with respect to absolute space. But it is hard to tell whether or how much the light speed (in space) is affected by the presence of air and the complicated motion of the earth in space. In reality, the direction and magnitude of the earth's absolute motion in space change always and irregularly due to the complicated motion of the earth in space.

In addition, I would point out one more fact that the speed of light checked in the MMX experiments was the proper speed of light in a given medium (air) and not the relative speed of light with respect to all observers. So it is wrong for relativists to use the MMX as the evidence of the invariance of light speed (relative speed) with respect to all observers.

The MMX experiments were not exact in principle. I suggest an improved interpretation of MMX in **Fig. 27-2**. Let us assume that the apparatus in **Fig. 27-2** is in uniform motion in space at the speed of v.

Before the apparatus starts moving, let us assume that the interferometer is stationary with respect to space. In this case, the monochromatic laser beam from source O proceeds until it reaches point

D on the half-silvered mirror and then splits in two. One of them follows the path of $D \to E \to D \to H$, and the path of $D \to G \to D \to H$. Let us call the former beam α and the latter β respectively.

The lengths of three interferometer arms are the same (L). That is, $DE = DG = DH = L$.

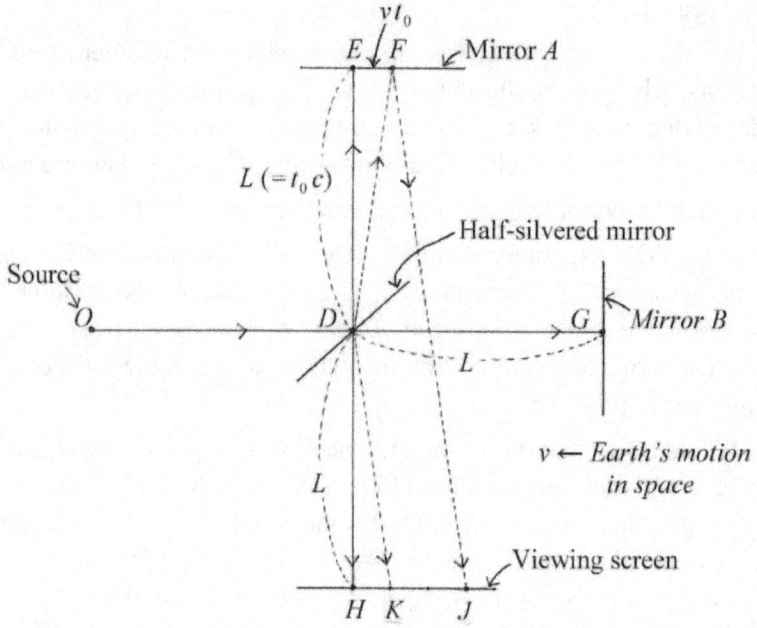

Fig. 27-2 Ahn's interpretation of the MMX

For convenience's sake, let us refer to the time any of the two beams takes to cover distance L as t_0. That is,

$$t_0 = \frac{L}{c}, \text{ or } L = t_0 c.$$

The total time beam α takes to cover the path of $DEDH$ is $3t_0$, and the total time beam β takes to cover the path of $DGDH$ is also $3t_0$. So in this case, there is only constructive interference on the viewing screen.

Next, let us assume that the interferometer is in uniform speed of v with respect to space in the direction indicated in **Fig. 27-3**. In this case the two laser beams will follow different paths as follows:

Special Relativity

Beam α: $D \rightarrow F \rightarrow J$
Beam β: $D \rightarrow G \rightarrow D \rightarrow K$.

Let us see find the times the two beams of light arrive at the viewing screen respectively.

The time beam α takes to travel distance DF is t_0. Likewise, we can see that the total time laser beam α takes to cover the path $D \rightarrow F \rightarrow J$ is $3 t_0$. (This is so regardless of the magnitude of the speed of the interferometer relative to space.)

The case of beam β is not that simple: The total time light beam β takes to cover the path $D \rightarrow G \rightarrow D \rightarrow K$ is

the time (t_1) beam β takes to travel from D to G plus

the time (t_2) beam β takes to travel from G to D plus

the time (t_3) beam β takes to travel from D to K.

Let us find t_1, t_2, and t_3 respectively.

We can find t_1 from **Fig. 27-3**.

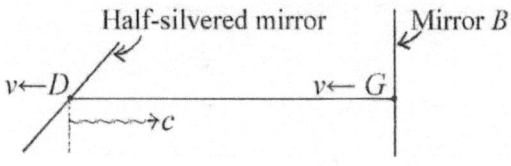

Fig. 27-3

Since D and G move to left at the speed v (see **Fig. 28-3**) and laser beam β travels from D to G at the speed c,

$c t_1 + v t_1 = $ distance DG. We know that $DG = L = t_0 c$.

Therefore, $t_1 = \dfrac{L}{c+v} = \dfrac{t_0 c}{c+v}$.

Note that "$c + v$", which is faster than light speed, is possible because this is a relative speed. (According to relativity, $c + v = c$. Then,

$t_1 = \dfrac{L}{c+v} = \dfrac{t_0 c}{c} = t_0$.)

We can find t_2 using **Fig. 27-4**.

In **Fig. 27-4**, D and G move to left at the same speed v, and beam β travels from G to D at the speed c. Therefore,

$vt_2 +$ distance $GD\ (= L) = ct_2$.

Therefore, $t_2 = \dfrac{L}{c-v} = \dfrac{t_0 c}{c-v}$.

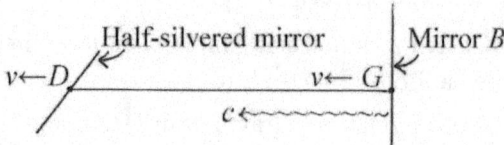

Fig. 27-4

(According to relativity, $c - v = c$. Then, $t_2 = \dfrac{L}{c-v} = \dfrac{t_0 c}{c} = t_0$.)

Finally, t_3 (the time beam β takes to travel from D to K) is t_0.

Therefore, the total time $t_1 + t_2 + t_3 =$

$$\dfrac{t_0 c}{c+v} + \dfrac{t_0 c}{c-v} + t_0 = \dfrac{3t_0 c^2 - t_0 v^2}{c^2 - v^2}.$$

The total time beam α takes to cover the path $D \to F \to J$ is $3t_0$ while the total time light beam β takes to cover the path $D \to G \to D \to K$ is $\dfrac{3t_0 c^2 - t_0 v^2}{c^2 - v^2}$.

Since $3t_0 = \dfrac{3t_0 c^2 - 3t_0 v^2}{c^2 - v^2}$, $3t_0 < \dfrac{3t_0 c^2 - t_0 v^2}{c^2 - v^2}$.

This means that beam α arrives at the viewing screen earlier than beam β does. And the spots the two beams get to are different from each other; beam α arrives at J, and beam β at K.

The fact that the two beams of light arrive at the viewing screen at different points and at different times means that there is no chance for the two beams to interference each other. This is so when the width of laser beam is narrow and the interferometer arms are long enough so that

Special Relativity

distance vt_0 (see **Fig. 27-2**) is conspicuous. But if the two split beams actually have considerable width and distance $v\,t_0$ is very small, interference phenomenon will occur. Then, will the interference constant in reality? Normally not so in reality because the motion/speed of the earth (and the interferometer on the earth) with respect to space is irregular. The presence of air (atmosphere) and the flow (wind) of air will also affect the interference irregularly.

Michelson and relativists interpret that the two beams arrive at the viewing screen on the same spot thought the arriving times are different from each other. **Fig. 27-2** shows that this interpretation is not correct.

After the MMX in 1887, Michelson and others repeated similar or improved version of MMX experiments, and the results were not complete "null effect." Yet relativists have ignored the "not complete null effect' to date.

I think that the presence of air, the turbulence of atmosphere (wind), and the irregular motion/speed of the earth with respect to space were the culprit of the irregular and "smaller-than-expected" degree of interference of the MMX experiments.

The motion of the earth is much more complicated than most people think. The earth's motion consists of the rotation and revolution motion of the earth, solar system, and the galaxy system, and the wobbling motion of the earth due to the presence of the moon. *The World Book Encyclopedia* states about the wobbling motion of the earth as follows: [3]

> The earth's moon has a diameter of about 2,160 miles (3,476 kilometers)—about a fourth that of the earth. The sun's gravity acts on the earth and the moon as if they are a single body with its center about 1,000 miles (1.600 kilometers) below the earth's surface. This spot is the earth-moon barycenter. It is the point of balance between the heavy earth and the lighter moon. The earth and the moon circle around the barycenter as they travel around the sun. The path of the barycenter around the sun is a smooth curve. The earth circles the barycenter and so follows a "wobbly" path around the sun.

Due to the irregular and uncontrollable air-current of the earth's atmosphere and the earth's complicated motion in space, it is wrong to expect any daily or seasonal pattern of the interference phenomenon of the MMX.

By the way, the concept of light speed Michelson maintained was of one dimension; he checked the light speed in the situations in which the observer or interferometer is aligned with the path of light. No scientists have thought of measuring the relative speed of light in 2-D or 3-D situations to date. But it is not easy to do so actually because we do not have faster-than-light means to check the speed of light in 2-D or 3-D situations. ♦

28

Time Dilation, Length Contraction, and Mass Increase Cannot Be Measured

Time dilation, length contraction/elongation, and mass increase do not occur. Even if any of these really occurs, there is no way for us to check that.

If we are to measure the length contraction effect of a moving body with a measuring rod, we have to move at the same speed of the object in order to measure the length of the target object. Then, the length of the measuring rod is reduced at the same rate as the target object (moving body) is. The case of measuring the time dilation or the mass increase is the same; a stationary observe cannot not confirm the time dilation effect or mass increase effect of a moving object.

So the job of measuring the effects of relativistic changes a kind of *rainbow catching game*: if we come close to a rainbow to get the pot of gold at the foot of the rainbow, the rainbow moves away. So we cannot get to the pot of gold forever. Relativists say that they can check the time dilation effect if a moving object makes a round trip and come back to the stationary observer. This is the story of the twin paradox. But the twin paradox is full with contradiction (see **Chapter 29**).

Einstein states in his train thought experiment as if we can measure the length contraction of a moving body with a ruler. [1] But in order to be able to measure the length contraction effect of a moving train with a ruler, the observer has to move at the same speed as the moving train. Then, the length of the ruler itself contracts at the same rate at which the length of the moving train contracts.

In the disc thought experiment, Einstein explains that the observer can confirm the length contraction of the circumference of the rim of the rotating disc using a "measuring rod" as follows: [1]

> If the observer [on the rotating disc] applies his standard measuring-rod (a rod which is short as compared with the radius of the disc) tangentially to the edge of the disc, then, as judged from the Galilean system [the inertial system relative to which the disc is rotating], the length of this rod will be less than 1, since moving bodies suffer a shortening in the direction of the motion. On the other hand, the measuring-rod will not experience a shortening in length, as judged from K [Galilean system], if it is applied to the disc in the direction of the radius. If, then, the observer first measures the circumference of the disc with his measuring-rod and then the diameter of the disc, on dividing the one by the other, he will not obtain as quotient the familiar number $\pi = 3.14...$, but a larger number, whereas of course, for a disc which is at rest with respect to K, this operation would yield π exactly.

In the above statement, Einstein makes several mistakes as follows:

Einstein thinks that only the rod, which is tangentially applied to the rim of the rotating disc, is experiencing the shortening in length whereas the rim of the rotating disc is not. Einstein is wrong here. If the measuring rod really gets shortened, the rim or the circumference of the disc should be shortened by the same rate as the measuring rod does. If we agree with Einstein's idea that the disc's radius, whose motion is always perpendicular to the direction of rotating motion of the disc, does not experience length contraction, then on dividing the shortened circumference by the un-shortened diameter, the observer will obtain a smaller, not larger, number than $\pi (= 3.14)$.

Another problem is that Einstein thought that the tangential speed of the rim of the disc as the relative speed of the rim with respect to *an observer* in the inertial Galilean space K. It is true that the tangential speed of the disc is uniform (constant) because the disc is in uniform rotation motion with respect to Galilean space K. This speed is a kind of proper speed which is independent of observers. That is, the tangential speed of the rim is **not** the relative speed with respect to an observer who rests in the Galilean space outside the rotating disc. The relative speed of

Special Relativity

a specific spot or a ruler on the rim of the disc with respect to a stationary observer in the Galilean system is not uniform but changes continuously and periodically as the disc rotates (see **Chapter 13**).

Fig. 28-1 is to illustrate Einstein opinion. The tangential speed of rod A is the angular speed (ω) times the radius (r). This speed is a kind of proper speed which is free of observers. This speed is not the relative speed of the rim or rod A with respect to an observer who rests outside the disc.

Einstein's disc thought experiment is important because it is the bridge that connects special relativity with general relativity. The disc thought experiment is also very important for dissenters because the fallacy of this experiment nullifies both special relativity and general relativity (see **Chapter 40**).

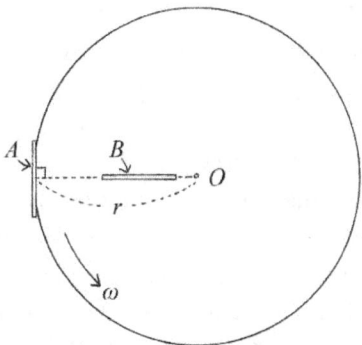

Fig. 28-1 The tangential speed of the circumference of the dis or rod A is independent of any observer outside the rotating disc.

Relativists say that particle accelerators have proved the mass increase effect of relativity. It is known that advanced particle accelerators can boost the speed of subatomic particles to the quasi-speed of light ($\approx c$). This word means, in terms of relativity, that the laboratory or accelerator or the earth itself moves the same speed (quasi-speed of light) toward the subatomic particles. Then, not only the accelerated particle but also the target and the accelerator and all other objects (including the scientists and heavenly bodies) will experience enormous rate of mass increase, length contraction, and time dilation. These stories are not science.

Physicists had discussed time dilation and length/distance contraction even before Einstein published special relativity. These physicists were ether-believers, and they did not discern proper speed from relative speed. Einstein simply borrowed or stole such existing theories and math for his relativity. [2], [3]

The original idea of length contraction was based on the existence of *material ether*. That is, if a body moves through the ether, the body would experience the resistance (pressure) of the ether and so the length of the body would be compressed in the direction of the motion of the body. This kind of idea would make sense if ether really exists. In this respect special relativity is based on ether theory. But Einstein negated the material ether. Consequently, his theory lost the ground or rationale of length contraction. If there is no ether, there will be no ether resistance/ pressure and no length contraction.

Einstein belatedly recognized the absoluteness of space and time because he realized that he could not explain the centrifugal force (which he had identified with gravity) without the absoluteness of space in his disc thought experiment. [4] Einstein's recognition of the absoluteness of space and time means that he gave up relativity in its entirety. Yet strangely enough, Einstein (and his followers as well) soon forgot what he said and maintained his former thought—the relativeness of space-time—until he died. Modern relativists ignore or forget (intentionally) what their master confessed in sobriety. In 1952, three years before he died, Einstein wrote in the *Fifth Appendix* of his fifteenth edition of *Relativity* that space is relative and freely movable (see **Chapter 18**). [5]

♦

29

Twin Paradox

Arthur Beiser explains the twin paradox in his textbook as follows (the related figure is omitted): [1]

> We are now in a position to understand the famous relativistic effect known as the twin paradox. This paradox involves two identical clocks, one which remains on the earth while the other is taken on a voyage into space at the speed of v and eventually is brought back. It is customary to replace the clocks with the pair of twins Dick and Jane, a substitution that is perfectly acceptable because the process of life—heartbeats, respiration, and so on—constitute biological clocks of reasonable regularity.
>
> Dick is 20 y old when he takes off on a space voyage at a speed of $0.80c$ to a star 20 light years away. To Jane, who stays behind, the pace of Dick's life is slower than hers by a factor of
>
> $$\sqrt{1-v^2/c^2} = \sqrt{1-(0.80c)^2/c^2} = 0.60 = 60\%.$$
>
> To Jane, Dick's heart beats only 3 times for every 5 beats of her beat; Dick takes only 3 breaths for every 5 of hers; Dick thinks only 3 thoughts for every 5 of hers. Finally Dick returns after 50 years have gone by according to Jane's calendar, but to Dick the trip has taken only 30 y. Dick is therefore 50 y old whereas Jane, the twin who stayed home, is 70 y old.
>
> Where is the paradox? If we consider the situation from the point of view of Dick in the spacecraft, Jane on the Earth is in motion relative to him at a speed of $0.80c$. Should not Jane be 50 y old when the spacecraft returns, while Dick is then 70—the precise opposite of what was concluded above?

> But the two situations are not equivalent. Dick changed from one inertial frame to a different one when he started out, when he reversed direction to head home, and when he landed on the Earth while Jane remained in the same inertial frame during Dick's whole voyage. The time dilation formula applies to Jane's observations of Dick, but not to Dick's observation of her.
>
> To look at Dick's voyage from his perspective, we must take into account that the distance L he covers is shortened to
>
> $$L = L_0\sqrt{1-v^2/c^2} = (20 \text{ light-years})\sqrt{1-(0.8c)^2/c^2} = 12$$
> light-years.
>
> To Dick, time goes by at the usual rate, but his voyage to the star has taken $L/v = 15$ y and his return voyage another 15 y, for a total of 30 y. Of course, Dick's life span has not been extended to him, because regardless of Jane's 50-y wait, he has spent only 30 years on the roundtrip.

Beiser's explanation of the differential aging of the twins is based on the assumption that the spacecraft, on which Dick is riding, maintains a uniform ***relative speed*** with respect to Jane who stays on the earth. But such an assumption is impossible in reality even if the spacecraft can maintain uniform proper speed with respect to space (with the help of gyroscopes) because of the complicated motion of the earth in space.

Even if we assume that Jane and Dick maintain a one-dimensional alignment throughout Dick's voyage, it is contradictory to tell which one of the two is in motion and which is not because relative motion is reciprocal. Beiser thinks he has solved this problem by saying that Dick is in non-uniform motion (accelerated motion) *when he started out, when he reversed direction to head home, and when he landed on the earth while Jane remained in an inertial frame*. Beiser's assumption that that the earth is an inertial frame is wrong. The earth is in complicated motion consists of rotation motion and revolution motion in space.

Even if we assume that the earth is an inertial system, it is wrong to insist on the unsymmetrical situation of Dick by saying that Dick was in accelerated motion *when he started out, when he reversed direction to head home, and when he landed on the earth while Jane remained in an inertial frame* because the portion (time) during which Dick was in accelerated motion was only a fraction of the whole trip. That is, during most time of the voyage, Dick was in uniform linear relative motion with respect to Jane (according to Beriser's assumption). If not, Beiser (and all other relativists) have no reason to apply time dilation factor or length contraction factor of $\sqrt{1-v^2/c^2}$ to the twin paradox. Therefore, Beiser's *asymmetrical situation theory* is not acceptable.

By the way, Beiser's asymmetrical theory is neutralized by Einstein's relativity of acceleration. Einstein asserted the relativity of acceleration his book *Relativity* (fifteenth ed.) as follows [2]: (The words in brackets are mine.)

> Even though it [chest] is being accelerated with respect to the "Galilean space [inertial frame]" first considered, we can nevertheless regard the chest as being at rest. We have thus good grounds for extending the principle of relativity to include bodies of reference which are accelerated with respect to each other, and as a result we have gained a powerful argument for a generalized postulate of relativity.

If Einstein's *relativity of acceleration* is true, Beiser's asymmetrical situation theory is of no use in rationalizing the different aging effect of the twin paradox. Of course Einstein' relativity of acceleration is wrong. Uniform motion in an inertial system is the state no additional force (energy) from outside is added or subtracted. But accelerated motion is the state continuous addition of force from outside is added or subtracted. Therefore, the state of being accelerated is not identical with the state of being at rest in an inertial system (see **Chapter 34**.) Nevertheless, Beiser keeps supporting the asymmetric aging effect as follows: [3]

> The nonsymmetric aging of the twins has been verified by experiments in which accurate clocks were taken on an airplane trip around the world and then compared with identical clocks that had been left behind. An observer who departs from an inertial system and then returns after moving relative to that system will always find his or her clocks slow compared with clocks that stayed in the system.

Beiser forgets that the airplane trip around the world was not a uniform relative motion relative to the clocks left behind. The airplane trip is a circular motion which is a kind of accelerated motion. Therefore the speed of the airplane relative to the clocks left behind was not uniform but changed continuously and periodically (see **Chapters 12** and **13**). The airplane experiment had nothing to do with the differential aging effect of the twin paradox. This sort of misconception of relativists comes from not distinguishing relative speed from proper speed.

If Beiser's statement concerning the differential time flow of the clocks in the airplane experiment was true, the probable cause might have been others than the special relativistic effect. For example, the asymmetrical condition of the temperature, air pressure, centrifugal effect, or other mechanical or physical condition(s) might have been the cause of the differential time flow of the clocks. I suggest Beiser and other relativists that they recheck the airplane experiment. I, for one, think that this kind of experiment is not worth doing.

By the way, *biological clock* is affected by temperature, pressure, nutrition, day/night light balance, and other physical or emotional condition rather than time dilation effect of relativity. If any asymmetrical aging effect of the twin paradox is tested in reality, it is much more likely that the twin who remains on the earth would remain much younger than the other twin who is taken on a voyage into space and then returned. Perhaps the latter would have been long dead by the time he/she gets returned home. ♦

Special Relativity

30

Muon Paradox

Relativists say that the distance a moving object covers is *shortened* (from the view point of an observer at rest). Relativists apply length contracting effect and time dilation effect to the twin paradox and the muon paradox differently. In the twin paradox, which we have discussed in the previous chapter, relativists apply distance contraction effect. But in the muon paradox, relativists apply the opposite effect—distance elongation. This is a huge problem.

But what is stranger is that relativists still argue that the muon paradox is the example of distance contraction effect. We need to scrutinize the story of the muon paradox in comparison with the twin paradox. Beiser explains the muon paradox in his textbook as follows: [1] (Figure number is reset for this book.)

LENGTH CONTRACTION
Faster means shorter

Measurement of length as well as of time intervals are affected by relative motion. The length L of an object in motion with respect to an observer always appears to the observer to be shorter than its length L_0 when it is at rest with respect to him. This contraction occurs only in the direction of the relative motion. The length L_0 of on object in its rest frame is called its **proper length**.

The length contraction can be derived in number of ways. Perhaps the simplest is based on time dilation and the principle of relativity. Let us consider what happens to unstable particles called muons that are created at high altitudes by fast cosmic-ray particles (largely protons) from space when they collide with atomic nuclei in the earth's atmosphere. A muon has a mass 207 times that of the electron and has a charge of either + e or − e; it decays into an electron or a positron after an average lifetime of 2.2 μs (2.2 × 10⁻⁶s).

Cosmic-ray muons have speeds of about 2.994 × 10⁸ m/s (0.998c) and reach sea level in profusion—one of them passes through each square centimeters of the earth's surface on the average slightly more often than once a minute. But in $t_0 = 2.2$ μs, their average lifetime, muons can travel a distance of only

$$vt_0 = (2.994 \times 10^8 \text{ m/s})(2.2 \times 10^{-6} \text{ s}) = 6.6 \times 10^2 \text{ m} = 0.66 \text{km}$$

Before decaying, whereas they are actually created at altitudes of 6km or more.

To solve this paradox, we note that the muon lifetime of $t_0 = 2.2$ μs is what an observer at rest with respect to a muon would find. Because the muons are hurtling toward us at the considerable speed of 0.998c, their lifetime are extended in our frame

$$t = \frac{t_0}{\sqrt{1-v^2/c^2}} = \frac{2.2 \times 10^{-6} \text{ s}}{\sqrt{1-(0.998)^2/c^2}} = 34.8 \times 10^{-6} \text{ s} = 34.8 \, \mu.$$

The moving muons have lifetime almost 16 times longer than those at rest. In a time interval of 34.8 μs, a muon whose speed is 0.998c can cover the distance

$$vt = (2.994 \times 10^8 \text{ m/s}) \times (34.8 \times 10^{-6} \text{s}) = 1.04 \times 10^4 \text{m} = 10.4 \text{ km}.$$

Although the lifetime is only $t_0 = 2.2$ μs in its own time of reference, a muon can reach ground from altitudes of as much as 10.4 km because in the frame in which these altitudes are measured, the muon lifetime is $t = 34.8$ μs.

What if somebody were to accompany a muon in its descent at v = 0.998c, so that to him or her the muon is at rest? The observer and the muon are now in the same frame of reference, and in this frame the muon's life time is only 2.2 μs. To the observer, the muon can travel only 0.66 km before decaying.

The only way to account for the arrival of the muon at ground level is if the distance it travels, from the point of view of an observer in the moving frame, is shorten by virtue of its motion. The principle of relativity tells us the extent of the shortening—it must be by the same factor of $\sqrt{1-v^2/c^2}$ that the muon lifetime

Special Relativity

is extended from the point of view of a stationary observer.

We therefore conclude that an altitude we on the ground find to be h_0 must appear in the muon's frame of reference as the lower altitude

$$h = h_0 \sqrt{1 - v^2/c^2} .$$

In our frame of reference the muon can travel h_0 = 10.4 km because of the time dilation (#11). In the muon's frame of reference, where there is no time dilation, the distance is abbreviated to

$$h = (10.4 \text{ km}) \sqrt{1 - (0.998c)^2/c^2} = 0.66 \text{ km}.$$

As we know, a muon traveling at $0.998c$ goes this far in 2.2 μs.

The relativistic shortening of distance is an example of the general contraction of lengths in the direction of motion:

Length contraction $L = L_0 \sqrt{1 - v^2/c^2}$.

Like time dilation, the length contraction is a reciprocal effect. To a person in a spacecraft, objects on the earth appear shorter than they did when he or she was on the ground by the same factor of $\sqrt{1 - v^2/c^2}$ that the space craft appears shorter to somebody at rest. The proper length L_o found in the rest frame is the maximum length any observer will measure. As mentioned earlier, only lengths in the direction of motion undergo contraction. Thus to an outside observer a spacecraft is shorter in flight than on the ground, but it is not narrower.

First of all, Beiser confuses relative speed with proper speed. The muon particles' speed 2.994×10^8 m/s ($0.998c$) is the proper speed that has nothing to do with observers. What relativists should deal with is *relative speed* and not proper speed. To the observers, who are not aligned with the line of motion of the muons, the relative speeds of the muon particles with respect to them (observers) are not constant; they vary continuously from $0.998c$ to zero (0). The relative speed becomes zero at the moment the observation angle is 90° (see **Chapters 4-8**). Since the speed 2.994×10^8 m/s ($0.998c$) of a muon paritcle is the proper speed, this speed is only for the observers who are aligned with the line of motion of the muon particle.

We should also note that the directions of the motion of muon particles are not all perpendicular to the surface of the earth. They fly in all direction due to the motion of the earth and the air currents of the atmosphere. Even if we assume that the direction of motion of a muon particle is perpendicular to the surface of the earth and that a human observer on earth is aligned with the line of motion of the muon particle, there are still other problems as follows:

Beiser states that the distance a moving muon particle covers is *elongated* (from the viewpoint of an observer on the earth). This story (distance elongation) is the opposite of the story of the twin paradox; In the twin paradox, Beiser explains that the distance a spacecraft covers is *contracted* (from the viewpoint of an observer on the earth).

Beiser explains that that the average lifetime of muon particles is 2.2 μs and that the muon particle can cover only 0.66 km during this average lifetime (when viewed from an imaginary observer who accompanies the muon particle). According to Beiser, the lifetime of a muon particle is **elongated** about 16 times (2.2 μs → 34.8 μs) by the factor of $\dfrac{1}{\sqrt{1-v^2/c^2}}$ when viewed from the observer on the ground. This is the effect of "getting older," which is opposite to the "stay younger effect" in the case of the twin paradox.

Distance Contraction Effect in the Twin Paradox

In the twin paradox, the original (proper) distance (20 light-years) between the earth and the star is contracted by the factor of $\sqrt{1-v^2/c^2}$ (= 0.6, where $v = 0.8c$) in the perspective of an observer (Jane) on the earth:

20 light-years (of proper distance) × $\sqrt{1-v^2/c^2}$ = 12 light years. This is length contraction (from the view point of the observer on the earth).

So, the time Dick takes to get to the star, according to Beiser, is 12 light-years ÷ 0.8c = 15years. During this time, 25 years pass for Jane on the earth (20light-years ÷ 0.8c = 25 y.) During the entire roundtrip to the star, it appears to Jane that Dick has earned 30 years of age (15 y × 2 = 30 y) whereas Jane on the earth earned 50 years of age (25y × 2 = 50 y).

Special Relativity

So the distance elongation effect and "getting older effect" in muon paradox are completely the opposite to the story of twin paradox. Yet Beiser still explains the muon paradox as the example of **length contraction** and **time dilation (aging less) effect**. Where does Beiser get this logic or trick? Beiser's trick, whether intended or not, is as follows:

The meaning of relativistic effect (change) is about how original (proper) distance L_0 is changed to L in the view point of an observer at rest. Distance 0.66 km is the original (proper) distance (L_0) a muon particle covers during its average life span; this distance is elongated to 10.4 km (L) when viewed from the observer who is at rest on the surface of the earth. That is,

$$\frac{L(elongated.length)}{L_0(proper.length)} = \frac{1}{\sqrt{1-v^2/c^2}} > 1.$$

(This is length elongation.)

However, Beiser's logic is reversed; he says that the proper distance is smaller than the elongated distance;

$$\text{Beiser's trick} = \frac{L_0(proper.length)}{L(elongaged.length)} = \sqrt{1-v^2/c^2} < 1.$$

(This is length contraction, according to Beiser's reversed logic.)

Beiser's explanation has other fundamental problems as follows:

Beiser (and all other relativists) pays attention only to the fact that the distance a muon particle covers is "elongated" and not the fact that the whole original (proper) altitude (6km) is also elongated at the same rate (about 16 times) when viewed from the observer on the earth. That is, the altitude is elongated as follows:

6km × 16 = 96km.

Why? According to special relativity, the entire straight line along which an object moves at the speed of v is elongated by the rate of

$$\frac{1}{\sqrt{1-v^2/c^2}}$$ (according to the muon paradox),

or contracted by the rate of $\sqrt{1-v^2/c^2}$ (according to the twin paradox).

Therefore, even if the distance a muon particle covers is elongated about 16 times (0.66m →10.4 km), this distance is way too short compared to the total elongated altitude (96km) (when viewed from the observer on the earth). So muon particles cannot reach the ground.

Now let us note that the diameter of a muon particle is elongated about 16 times when viewed by an observer on the earth. By the way, relative speed and relativistic effects are reciprocal. In the view point of a muon, the muon itself is stationary while all the objects (including the earth and many other heavenly bodies) that are aligned with the line of motion of the muon are in motion at the speed of $0.998c$ with respect to the muon. Then it should be true that not only the height of an observer on the earth but also the diameter of the earth is elongated about 16 times in the direction of the earth's motion toward the muon. These are the case when we apply distance elongation as Beiser do to the muon paradox [see the diagram (b) in **Fig. 30-2**]. However, according to the logic of twin paradox, the reverse should be true. That is, the distance the earth covers and the diameter of the earth are reduced to the rate of 1/16 [see diagram (a) in **Fig. 30-2**].

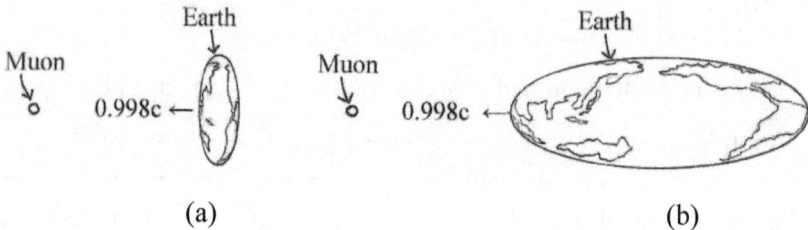

Fig. 30-2 Contraction effect (a) and elongation effect (b)

If length elongation effect is true, not only the diameter of the earth but also everything else on the earth, including private members of men, will be elongated about 16 times if these are aligned with the line of motion of a muon particle.

Relativists may say that muon particles are so small and light compared to that of earth so that the diameter elongation effect happens only to the diameter of the muon and not to that of the earth. This logic is untrue. Relativistic effects are reciprocal and have to do with relative speed between two involved objects and not with the ratio of mass or size of the two.

Special Relativity

The motion/speed of a single muon particle will not affect only the diameter/mass of the earth but also all other the heavenly bodies (including galaxies) which are happen to be aligned with the line of motion of a muon particle.

Imagine that the motion of a single particle has the power to increase (or decrease) the diameters and masses of all the objects that are aligned with the line of motion of the objects. By the way, we have learned that relative speed occurs not only in 1-D situations but also in 2-D and 3-D situations. The motion of a single muon really affects (increases or decreases) the size and mass of the entire universe?

What if we apply photons instead of muons? If we do, the diameter of the earth (and any other heavenly body) will be zero (0) (if we apply distance contraction effect) or infinite (∞) (if we apply distance elongation effect). The mass increase or decrease will occur at the same rate. What a magic! The true color of relativity is magic; it started with a magical idea.

Will relativists argue unanimously that we cannot replace muons with photons in the discussion on relativistic changes because photons have no rest mass? How small mass is enough not to be applied to relativity? The difference between zero mas and yes-mass is infinitely small.

The expression "$\lim_{m \to 0} m$" shows that the boundary between mass and no-mass is practically zero (0).

Again, relativity is about relative speed and not about the ratio of masses. Relativists should not be discriminatory; the motion of a muon is not different from that of a photon. Motion is motion and speed is speed regardless of the magnitude of the mass of an object. The law of physics should the same for all objects/phenomena, according to the postulation of Einstein.

By the way, if we apply $c \pm v = c$, the Lorentz factor becomes 1.

$$\sqrt{1 - v^2/c^2} = \sqrt{\frac{(c+v)(c-v)}{c^2}} = \sqrt{\frac{c \times c}{c^2}} = 1. \text{ (See \textbf{Chapter 24}.)} \blacklozenge$$

31

Clock Paradox

Beiser explains time dilation effect using two imaginary light clocks in his text book as follows: [1] (The numbers of figures are reset for this book. The equation numbers (1), (2), etc. are re-given for this chapter only. The words in the brackets are mine.)

TIME DILATION
A moving clock ticks more slowly than a clock at rest

Measurements of time intervals are affected by relative motion between an observer and what is observed. As a result, a clock that moves with respect to an observer ticks more slowly than it does without such motion, and all processes (including those of life) occur more slowly to an observer when they take place in a different inertial frame.

If someone in a spacecraft finds that the time interval between two events in the spacecraft is t_0, we on the ground would find that the time interval has the longer duration t. The *quantity* t_0, which is determined by events that occur *at the same place* in an observer's frame of reference, is called the **proper time** of the interval between the events. When witnessed from the ground, the events that mark the beginning and end of the time interval occur at different places, and in consequence the duration of the interval appears longer than the proper time. This effect is called **time dilation** (to dilate is to become larger).

To see how time dilation comes about, let us consider two clocks, both of the particularly simple kind shown in **Fig. 31-3**. In each clock a pulse of light is reflected back and forth between two

Special Relativity

mirrors L_0 apart. Whenever the light strikes the lower mirror, and electric signal is produced that marks the recording tape. Each mark corresponds to the tick of an ordinary clock.

One clock is at rest in a laboratory on the ground and the other is in a spacecraft that moves at the speed of v relative to the ground. An observer in the laboratory watches both clocks: does she find that they tick at the same rate?

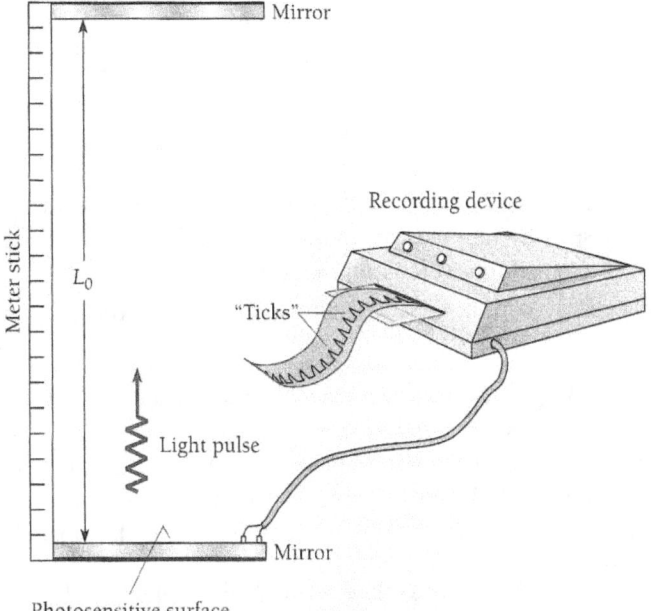

Fig. 31-1 A simple clock. Each "tick" corresponds to a round trip of the light pulse from the lower mirror to the upper one and back.

Figure 31-2 shows the laboratory clock in operation. The time interval between ticks is the proper time t_0 and the time needed for the light pulse to travel between the mirrors at the speed of light c is $\dfrac{t_0}{2}$.

Hence $\dfrac{t_0}{2} = \dfrac{L_0}{c}$, and

211

$$t_0 = \frac{2L_0}{c} \qquad \text{(1)}$$

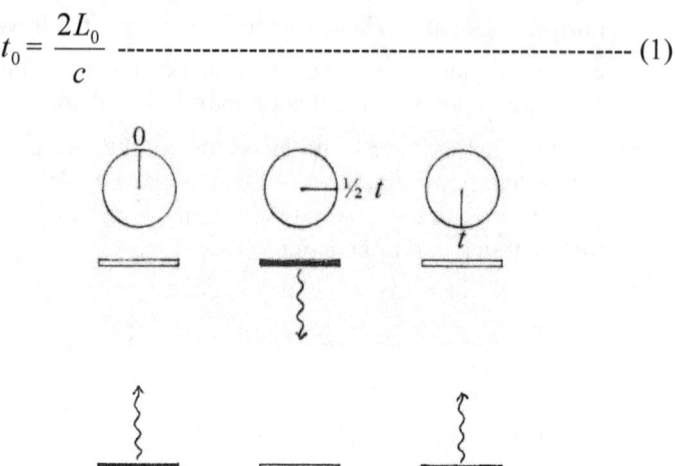

Fig. 31-2 A light-pulse clock at rest on the ground as seen by an observer on the ground. The dial represents a conventional clock on the ground.

[Correction: t in the figure should be t_0 / Ahn.]

Fig. 31-3 shows the moving clock with its mirrors perpendicular [*parallel?*] to the direction of motion relative to the ground. The time interval between ticks is t. Because the clock is moving, the light pulse, as seen from the ground, follows a zigzag path. On its way from the lower mirror to the upper one in the time $\frac{t}{2}$, the pulse travels a horizontal distance of $v(\frac{t}{2})$ and a total distance of $c(\frac{t}{2})$. Since L_0 is the vertical distance between the mirrors,

$$\left(\frac{ct}{2}\right)^2 = L^2{}_0 + \left(\frac{vt}{2}\right)^2$$

$$\frac{t^2}{4}(c^2 - v^2) = L^2{}_0$$

$$t^2 = \frac{4L^2{}_0}{c^2 - v^2} = \frac{(2L_0)^2}{c^2(1 - v^2/c^2)}$$

Special Relativity

$$t = \frac{2L_0/c}{\sqrt{1-v^2/c}} \qquad (2)$$

But $2L_0/c$ is the time interval t_0 between ticks on the clock on the ground, as in **Eq. (1)**, and so

$$\text{time dilation } t = \frac{t_0}{\sqrt{1-v^2/c^2}}. \qquad (3)$$

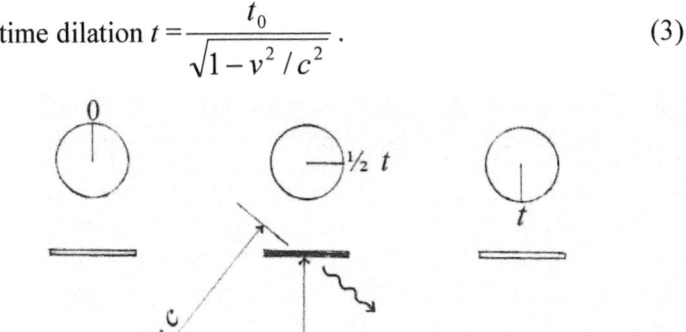

Fig. 31-3 A light clock in a spacecraft as seen by an observer on the ground. The mirrors are parallel to the direction of motion of the spacecraft. The dial represents a conventional clock on the ground.

Here is a reminder of what the symbols in **Eq. (3)** represent:

t_0 = time interval on clock at rest relative to an observer = proper time

t = time interval on clock in motion relative to an observer

v = speed of relative motion

c = speed of light

Because the quantity $\sqrt{1-v^2/c^2}$ is always smaller than 1 for a moving object, t is always greater than t_0. The moving clock in the spacecraft appears to tick at a slower rate than the stationary one on the ground, as seen by an observer on the ground.

Exactly the same analysis holds for measurement of the clock on the ground by the pilot of the spacecraft. To him, the light pulse of the ground clock follows a zigzag path that requires a total time *t* per round trip. His own clock, at rest in the spacecraft, ticks at intervals of t_0. He too finds that

$$t = \frac{t_0}{\sqrt{1 - v^2/c^2}}$$

so the effect is reciprocal: *every* observer finds that clocks in motion relative to him tick more slowly than clocks at rest relative to him.

Our discussion has been based on a somewhat unusual clock. Do the same conclusions apply to ordinary clocks that use machinery – spring-controlled escapements, tuning forks, vibrating quartz crystals, or whatever – to produce ticks at constant time intervals? The answer must be yes, since if a mirror clock and a conventional clock in the spacecraft agree with each other on the ground but not when in flight, the disagreement between them could be used to find the speed of the spacecraft independently of any outside frame of reference–which contradicts the principle that all motion is relative.

COUNTERPROOF

Firstly, I would indicate two fundamental problems with Beister's light clock experiment:

The first one is that the laboratory on the ground is not an inertial system because of the complicated motion of the earth in space. Therefore, even if we concede that the spacecraft maintains a uniform velocity (speed + direction) in space with the help of gyroscopes, it is wrong for relativists to think that the spacecraft and the laboratory on the ground can maintain a uniform relative (see **Chapter 29**).

The second problem is that the speed (proper speed) of light is not constant with respect any clocks or laboratories. The speed (proper speed) of light is constant with respect to absolute space (see **Chapter 21**).

Beiser, as well as all other relativists including Einstein, think that any observers or laboratories are inertial frame of reference (regardless of the types of motion of these) and that the speed (relative speed) of light is

Special Relativity

constant with respect to any observers or laboratories. Relativists are wrong.

There are other minor problems; Beiser states, "An observer in the laboratory watches both clocks [one in the laboratory on the ground and the other on the spacecraft]: does she find that they tick at the same rate?" Even if we concede that the clock on the moving spacecraft ticks slower than the clock on the ground, it is impossible for the observer on the ground to check whether the clocks tick at different rate. The reason is as follows:

If one is to check the ticking rate (time flow rate) of a clock, the information of the clock has to reach the ear or eye (retina) of the observer. So it takes some time for such information to travel (at the speed of light or sound) from the clock to the observer. So there occurs a "time lag." If the distance of the clock from the observer is very close as in the case of the clock and observer on the ground we may ignore such a time lag which is very small. But when a clock is on a moving spacecraft in pace and an observer is on the ground, the "time lag" will be considerable and ever-growing. For these reasons, it is not possible for an observer on the ground to compare the time flow rates at then-and-there basis. Bu this problem is minor one compared to the two fundamental problems aforementioned.

Beisers' light clock is imaginary one; it does not work in reality. But I do not object thought experiments themselves as long as they are based on logic and analysis.

Even if we assume that the earth (and the observer on the ground) is inertial state or stationary with respect to space and the spacecraft maintains a uniform motion with respect to the observer on the ground, time dilation does not occur as Beiser or other relativists expect. I will prove this in the following thought experiment with a light clock whose construction is a bit different from Beiser's (see **Fig. 31-4**). I would call this experiment *Ahn's light clock experiment*.

Ahn's light clock is so made that a photon (P) reflected back and forth two mirrors (Mirror-*A* and Mirror-*B*) L distant apart. Time-recording device is conveniently tucked under the panel of the light clock so that it is not showing in the diagram. Each tick is recorded every time photon *P* hits any of the two mirrors.

Harvard Physicists Confuse Relative Speed with Proper Speed

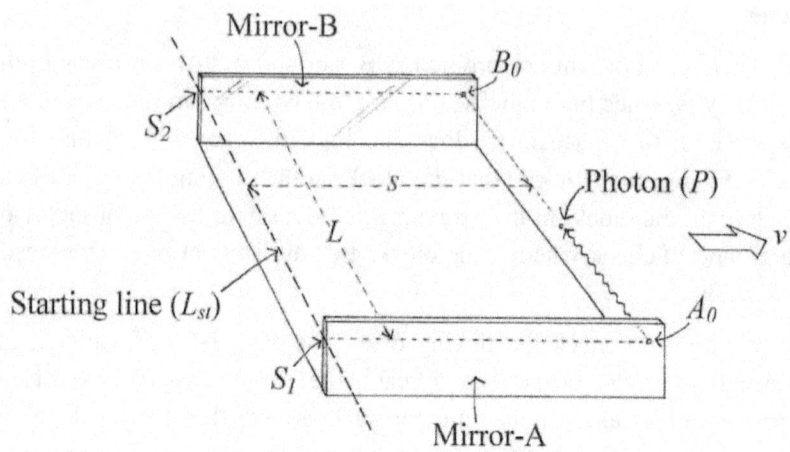

Fig. 31-4 Ahn's light clock experiment.
(Time recording device is omitted.)

Assume that Ahn's clock is stationary with respect to absolute space. It starts moving from starting line L_{st} at time $t = 0$ (zero) at the uniform velocity v in the direction indicated by an arrow. The direction of motion of Ahn's clock is *perpendicular* to the starting line, and the moving direction of photon P is parallel to the starting line.

If the clock does not move and stays at starting line L_{st}, photon P will travel forth and back between two points—A_0 (on Mirror-A) and B_0 (on Mirror-B) [see diagram (a) in **Fig. 31-5**]. The time photon P takes to travel from A_0 and B_0 is $t = \frac{L_0}{c}$. Both A_0 and B_0 are distance s from starting line L_{st}. That is, $S1A0 = S_2B_0 = L$. The direction of motion of photon P is *parallel* to start line L_{st};

Now, if the light clock starts moving at the speed v in the direction of the arrow, the following will happen:

At time $t = 0$ [see diagram (a) in **Fig. 31-5**], photon P is at point A_0 on Mirror-1. At this moment, photon P is distance s from starting line L_{st}. That is, $A_0S_1 = s$.

216

Special Relativity

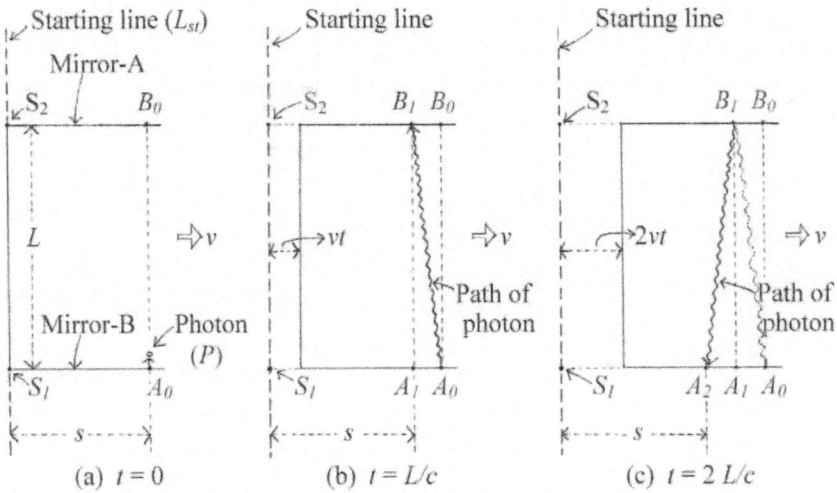

Fig. 31-5 The position and path of photon P at $t = 0$, $t = L_0/c$, and $t = 2 L_0/c$.

At time $t = L/c$ [see diagram (b)], photon P arrives at point B_1 (instead of B_0) on Mirror-2 because the light clock moves at the speed of v perpendicularly to starting line L_{st}. At this moment $B_1 B_0 = vt$ and photon P is distance s from starting line L_{st}. That is, $S_2 B_1 = s$.

We should note that photon P moves at the speed c parallel to starting line L_{st}. The speed of photon P has no "perpendicular component" with respect to starting line L_{st}. Therefore the distance photon P with respect to L_{st} is always s and does not change. This is the main difference of Ahn's light clock experiment compared to relativists' light clock experiment.

Relativists *customarily* think that the speed of photon P has a "perpendicular component of v as that of the speed of the light clock. This opinion is based on the idea (wrong idea) that photon P has inertia so that P moves at the same speed (v) and same direction of the light clock. Photons have no inertia; photon P moves parallel to starting line L_{st}. *irrespective* of the speed and direction of the light clock. According to this phenomenon, the following statement is true:

At time $t = 2L/c$, (see diagram (c) in **Fig. 31-5**], photon P arrives at point A_2 on Mirror-A. At this moment, $A_0 A_1 = A_2 A_1 = vt$, and $A_0 A_2 = 2vt$. Photon P is still distance s from starting line L_{st}. That is, $S_1 A_2 = s$.

The distance photon P covers for each time t is L and does not change. Therefore, there is no time dilation effect at all regardless of the dimension of the speed (v) of the light clock in space. ♦

32

$E = mc^2$

Einstein's energy equation $E = mc^2$ is known to all as the highlight of relativity. Wikipedia explains this energy equation as follows: [1] (The figure number is given to fit this book.)

Mass–velocity relationship

In developing special relativity, Einstein found that the kinetic energy of a moving body is

$$E_k = m_0(\gamma - 1)c^2 = \frac{m_0 c^2}{\sqrt{1 - v^2/c^2}} - m_0 c^2,$$

with v the velocity, m_0 the rest mass, and γ the Lorentz factor ($\frac{1}{\sqrt{1 - v^2/c^2}}$). He included the second term on the right to make sure that for small velocities, the energy would be the same as in classical mechanics:

$$E_K = \frac{1}{2} m_0 v^2 + \dots.$$

Without this second term, there would be an additional contribution in the energy when the particle is not moving.

Einstein found that the total momentum of a moving particle is:

$$P = \frac{m_0 v}{\sqrt{1 - v^2/c^2}}.$$

and it is this quantity which is conserved in collisions. The ratio of the momentum to the velocity is the relativistic mass, m.

$$m = \frac{m_0}{\sqrt{1 - v^2/c^2}}$$

And the relativistic mass and the relativistic kinetic energy are related by the formula:

$$E_K = mc^2 - m_0 c^2.$$

Einstein wanted to omit the unnatural second term on the right-hand side, whose only purpose is to make the energy at rest zero, and to declare that the particle has a total energy which obeys:

$$E = mc^2$$

which is a sum of the rest energy $m_0 c^2$ and the kinetic energy. This total energy is mathematically more elegant, and fits better with the momentum in relativity. But to come to this conclusion, Einstein needed to think carefully about collisions. This expression for the energy implied that matter at rest has a huge amount of energy, and it is not clear whether this energy is physically real, or just a mathematical artifact with no physical meaning.

In a collision process where all the rest-masses are the same at the beginning as at the end, either expression for the energy is conserved. The two expressions only differ by a constant which is the same at the beginning and at the end of the collision. Still, by analyzing the situation where particles are thrown off a heavy central particle, it is easy to see that the inertia of the central particle is reduced by the total energy emitted. This allowed Einstein to conclude that the inertia of a heavy particle is increased or diminished according to the energy it absorbs or emits.

We have seen in the previous chapters the following:
- Einstein did not distinguish relative speed from proper speed; he mistook proper speed for relative speed (see **Chapter 7**).
- Lorentz transformation is wrong (see **Chapter 24**).
- The isotropy of light speed is wrong (see **Chapter 17**).
- Einstein's knowledge of space and time is wrong (see **Chapter 18**).

Special Relativity

- Einstein's knowledge of inertial motion is wrong (see **Chapter 16**).
- Special relativity violates the conservation of mass-energy (see **Chapter 33**).

Therefore, the energy equation is wrong.

Then, what is the secret of atomic bomb? Most people think that atomic bombs could be made by Einstein's energy equation. This is a myth. It is true that nuclear fission releases tremendous energy. But nuclear bombs have nothing to do with relativity.

The fact that nuclear fission releases lot of energy was discovered by other scientists independently of Einstein's energy equation. According to *Max Plank Institute for Gravitational Physics* (formerly, Albert Einstein Institute), Einstein's political role, rather than his energy equation, helped manufacture the atomic bomb. **Einstein Online**, the official internet site of the *Max Plank Institute for Gravitational Physics*, explains the story of the energy equation and atomic bomb as follows: [2] (Figure number is given for this book.)

From $E = mc^2$ to the atomic bomb

When Einstein's most famous formula $E = mc^2$ is mentioned, the atomic bomb is usually not far behind. Indeed there is a connection between the two, but it is subtle, and sadly, some popular159 science texts get it wrong: they will tell you that a nuclear explosion is "caused by the transformation of matter and energy" according to Einstein's formula, and that the gigantic conversion factor c^2 is responsible for the immense power of such weapons.

But first things first. Let's have a look at what Einstein really did say about the relation between mass and energy.

Equivalence or transformation?

For Einstein, mass (more precisely: relativistic mass; the property that determines how difficult it is to change a body's speed or its direction of motion) and energy are simply two different names for one and the same physical quantity. Whenever a system has an energy E, it automatically has the relativistic mass $m = E/c^2$; whenever a system has the mass m, you need to assign it an energy $E = mc^2$. Once the mass is known, so is the energy, and vice versa. In that context, it makes no sense to talk about the "transformation of mass into energy"- where there's one, there's the other.

The context in which "transformation of mass into energy" does make sense is a bit different. It is intimately connected with the fact that there are different kinds of energy. Already in classical, pre-Einstein physics, the concept of "energy" comprises a plethora of sub-definitions for different sorts of energy, sub-definitions like those for the kinetic energy associated with any moving body, the energy of electromagnetic radiation, thermal energy or the binding energy that needs to be taken into account whenever there is a force holding together two objects to form a composite object. Yet all these different definitions can be viewed as facets of a single physical quantity, energy. The reason is the possibility of transformations between the different energy forms. For instance, you can increase a body's temperature (and thus its thermal energy) by letting it absorb electromagnetic radiation energy. In these transformations, the total sum of all the different kinds of energy - the total energy - is constant over time. Energy can be transformed from one variety into another, but it can neither vanish nor be created from nothing.

Fig. 34-1 Ten seconds after the ignition of the first atomic bomb, New Mexico, July 16, 1945.
[Source: Los Alamos National Laboratory]

A new kind of energy

This *conservation of energy* holds not only in classical

physics, but also in special relativity. However, in relativity, the definitions of the different species of energy are a bit different and, most importantly, there is a completely new type of energy: even if a particle is neither moving nor part of a bound system, it has an associated energy, simply because of its mass. This is called the particle's rest energy, and it is related to the particle's rest mass as

$$\text{rest energy} = (\text{rest mass}) \cdot c^2.$$

Compared with other types of energy, rest energy is very much concentrated. For example: If you use a television tube to accelerate an electron to 20,000 kilometers per second, the kinetic energy gained is still only about five hundred times smaller than the electron's rest energy. Also, this rest energy is about a hundred times larger than the radiation energy of a high-energy X-ray photon. This high concentration is important for processes where rest energy (or, equivalently, rest mass) is converted to more common forms of energy. For instance, when a particle and its antiparticle annihilate and vanish in a puff of electromagnetic radiation, comparatively little matter is transformed into rather a lot radiation.

Studying the masses of different types of atomic nuclei, you will find that in nuclear fission - the process that powers an ordinary atomic bomb -, some "nuclear rest energy" or "nuclear rest mass" is transformed into other forms of energy. For example, the rest mass of a nucleus of uranium-235 is slightly larger than the combined rest masses of the nuclear fragments into which it splits during nuclear fission. Here's where $E = mc^2$ comes into play: This mass difference corresponds to the energy set free during nuclear fission. So is it, after all, true that Einstein's formula explains the power of the nuclear bomb - and that the large conversion factor c^2 is responsible for the immense amounts of energy released?

Binding energies: nuclei vs. molecules

Not at all. Different process, same calculation: For chemical reactions, there are tiny mass differences as well. To pick an example: When hydrogen and oxygen explosively combine to make water, the sum of the rest masses of the initial hydrogen and oxygen atoms is just a little bit less than the sum of the rest masses of the resulting water molecules. The same is true for the chemical reactions involving spontaneous oxidation - in other words: burning. The same formula applies: The mass difference, multiplied by c^2, gives the energy set free during the chemical reaction. Same formula, same conversion factor - yet chemical reactions are much less violent than nuclear explosions. This clearly shows that the difference between nuclear and chemical reactions must be due to something other than $E = mc^2$.

To see where the difference lies, one must take a closer look. Atomic nuclei aren't elementary and indivisible. They have component parts, namely protons and neutrons. In order to understand nuclear fission (or fusion), it is necessary to examine the bonds between these components. First of all, there are the nuclear forces binding protons and neutrons together. Then, there are further forces, for instance the electric force with which all the protons repel each other due to the fact they all carry the same electric charge. Associated with all of these forces are what is called binding energies - the energies you need to supply to pry apart an assemblage of protons and neutrons, or to overcome the electric repulsion between two protons. (More information about these binding energies and their role in nuclear fission and fusion can be found in the spotlight topic *Is the whole the sum of its parts?*)

Only with the systematics of these forces and binding energies well understood were physicists able to uncover the laws behind nuclear fission and fusion: The strength of the nuclear bond depends on the number of neutrons and protons involved. It varies in such a way that binding energy is released both in splitting up a heavy nucleus into smaller parts and in fusing light nuclei into heavier ones. This, as well as the chain reaction phenomenon, explains the immense power of nuclear bombs.

Einstein's formula plays second fiddle in that derivation - it's all about different kinds of energy. Sure, there are some radioactive decay processes following nuclear fission, and, if so inclined, one can view the decay of a neutron decaying into a slightly lighter proton as a transformation of rest energy into other energy forms. But these additional processes contribute a mere 10 per cent of the total energy set free in nuclear fission. The main contribution is due to binding energy being converted to other forms of energy - a consequence not of Einstein's formula, but of the fact that nuclear forces are comparatively strong, and that certain lighter nuclei are much more strongly bound than certain more massive nuclei.

Still, $E = mc^2$ had a supporting role in the story of nuclear fission research. Not as the mechanism behind nuclear power, but

Special Relativity

as a tool: Because energy and mass are equivalent, highly sensitive measurements of the masses of different atomic nuclei gave the researchers important clues about the strength of the nuclear bond. Einstein's formula does not tell us *why* the nuclear binding energies are as large as they are, but it opens up one way (among several) to measure these binding energies. (More about this application of Einstein's formula can be found in the spotlight topic *Is the whole the sum of its parts?*)

In fact, Einstein's politics played a more decisive role in the story of the atomic bomb than his physics. Following a request by the physicist Leo Szilard, Einstein wrote a letter to president Roosevelt, explaining about the potential power of nuclear weapons and the possibility of Nazi Germany developing such weapons, and urging the president to take action. Einstein's letter played its part in setting into motion the political process that culminated in the Manhattan project—the development, construction and testing of the first nuclear bombs.

I am not in the position to judge or comment about internal energy. But I do not think the factor c^2 has any meaning because I do not agree with the concept of c, which most physicists think as the same for all observers. c is neither proper speed nor relative speed. There is no such speed c that is the same for any/all observers. ♦

33

The Law of Conservation of Matter and Energy

Einstein stated that the mass of an object is increased if it moves (with respect to an inertial system), that the increased mass equals to the energy that causes the motion of the object, and that the increased mass is hardly detected only unless the speed of the object is not very small as compared with the velocity of light. Einstein called it the conservation of mass-energy. Einstein explains the conservation of energy in his book *Relativity* as follows: [1]

> It is clear from our previous considerations that the (special) theory of relativity has grown out of electrodynamics and optics. In these fields it has not appreciably altered the predictions of theory, but it has considerably simplified the theoretical structure, *i.e.* the derivation of laws, and—what is comparably more important—it has considerably reduced the number of independent hypothesis forming the basis of the theory. The special theory of relativity has rendered the Maxwell-Lorentz theory so plausible, that the latter would have been in generally accepted by physicists even if experiment had decided less unequivocally in its favour.
>
> Classical mechanics required to be modified before it could come into line with the demands of the special theory of relativity. For the main part, however, this modification affects only the laws for rapid motions, in which the velocities of matter v are not very small as compared with the velocity of light. We have experience of such rapid motions only in the case of electrons and ions; for other motions the variations for the laws of classical mechanics are too small to make themselves evident in practice. We shall not consider the motion of stars until we come to speak of the general

Special Relativity

theory of relativity. In accordance with the theory of relativity the kinetic energy of a material point of mass m is no longer given by the well-known expression $m\dfrac{v^2}{2}$ but by $\dfrac{mc^2}{\sqrt{1-\dfrac{v^2}{c^2}}}$.

This expression approaches infinity as the velocity v approaches the velocity of light c. The velocity must therefore always remain less than c, however great may be the energies used to produce the acceleration. ....

The most important result of a general character to which the special theory of relativity has led is concerned with the conception of mass. Before the advent of relativity, physics recognized two conservation laws of fundamental importance, namely, the law of conservation of energy and the law of conservation of mass; these two fundamental laws appeared to be quite independent of each other. By means of the theory of relativity they have been united into one law, and what meaning is to be attached to it.

The principle of relativity requires that the law of the conservation of energy should not only with reference to a co-ordinate system K, but also with respect to every co-ordinate system K' which is in a state of uniform motion of translation relative to K, or, briefly, relative to every "Galilean" system of co-ordinate. In contrast to classical mechanics, the Lorentz transformation is the deciding factor in the transition from one such system to another.

Einstein's law of conservation of energy and matter has flaws as follows:

Einstein unified electrodynamics, optics, and classical mechanics into one by using Lorentz transformation. Electrodynamics and classical mechanics are entirely different entities from each other; these are not the things to be unified by any equations. Lorentz transformation is based on "ungrounded postulate" that the speed of light is the same for all observers. Lorentz and Einstein did not distinguish relative speed from proper speed (see **Chapters 7, 17, 21**.)

Einstein's concept of inertial co-ordinate system is incorrect. He thought not only a system (environment) but also any object or observer can be inertial co-ordinate system regardless of the type of motion—uniform motion or accelerated motion (see **Chapter 16**). Einstein's concept of mass is wrong; he identified inertial mass with gravitational mass (see **Chapter 35**). Einstein's *relativity of acceleration* is wrong (see **Chapter 34**).

The argument that the increase of speed contribute to the increase of mass might be plausible when it comes to ***actual motion/speed*** (see the concept of actual motion/speed in **Chapter 1**). But Einstein did not distinguish actual motion/speed from ***virtual motion/speed***. When it comes to relative motion, if one party is in actual motion, the other party may be in virtual motion. According to special relativity, mass increase occurs not only to the object in actual motion but also to the object in virtual motion. This means that, according to relativity, mass increase occurs to the object that is stationary in a given inertial system if its counterpart object is in actual motion.

Einstein thought that relative motion/speed occurs only in 1-situations. But relative motion/speed occurs not only in 1-D situations but also in 2-D and 3-D situations. This means that if an object is in actual motion, all other objects in the universe are in relative or virtual motion though the magnitudes of relative or virtual speeds are not the same. Do the masses of all the objects in the universe increase if an object is in actual motion? I do not think so since special theory is based on many fundamental fallacies. Relativity (both special relativity and general relativity) is a bundle of misconceptions.

The law of conservation of mass and energy means that the total mass and energy in the universe is of certain absolute quantity. But the core of relativity is that everything, including mass and energy, is relative. I wonder where Einstein got the idea of the conservation of energy and mass.

There are many different types of energies—kinetic energy, potential energy, bonding energy, electro-magnetic energy, gravitational energy, etc. I do not know whether all these types of energies are the same thing or not. I suspect that energy and matter are different entities from each other. ♦

Part II

General Relativity

Inertial force, which is pseudo force in substance, is the resistance against the change of speed or direction of motion. Gravity is one of the four real forces that exist in nature (see **Chapter 36**). The curving of light beam in the Einstein's chest thought experiment (see **Chapter 37**) is simply the phenomenon of Newtonian relativity due to the accelerated motion of the chest.

Relativists say that inertial force and gravity are so alike each other that there is no way to tell them apart. But I have found at least eight different characteristics that clearly distinguish inertial force from gravity. Inertial force is one thing and gravity is totally another (see **Chapter 36**).

34

"Relativity of Acceleration" Is Wrong

Einstein states in his chest thought experiment that the state of being accelerated can be regarded as being at rest as follows: [1] (The words in the brackets are mine.)

> Even though it [a chest] is being accelerated with respect to the 'Galilean space [inertial frame of reference]' first considered we can nevertheless regard the chest as being at rest. We have thus good grounds for extending the principle of relativity to include bodies of reference which are accelerated with respect to each other.

I would refer to the above assertion of Einstein as *the relativity of acceleration*. But I disagree with this kind of relativity. The reason is as follows:

Suppose that train A is stationary on a straight railroad and that train B on the same railroad is moving away from train A at a uniform acceleration, say a. (We assume that the embankment on which the two trains are situated is flat and in inertial state.) If we pay attention only to the two trains and not the given environment (embankment), it is true that two trains are in accelerated motion with each other. But there is clear difference if we check which one is in actual accelerated motion with respect to the given environment and which is not. Train B is in *actual accelerated motion* with respect to the given environment (embankment) whereas train A is in **virtual accelerated motion** with respect to train B. Train A is actually stationary with respect to the given system. Train A is only in virtual accelerated motion with respect to train *B* (see the definitions of actual motion and virtual motion in **Chapter 1**).

In the above case, jerk or inertial force, which Einstein misidentified with gravity, occurs only inside train *B*, which is in actual accelerated motion, and not in train *A*. Actual accelerated motion is the state in which uniform force is continuously added or subtracted whereas the state of being at rest is the state in which no such force is added or subtracted.

In reality, embankment is neither flat nor in inertial state. This is because of the round shape of the earth and the complicated motion of the earth in space. Einstein considered any train, chest, embankment or even the earth as inertial coordinate system. Einstein was wrong. Einstein believed that space is relative and that all motion is relative. Therefore he thought that any kind of system (environment) can be an inertial system regardless of its state or type of motion.

I would say there is no true inertial environment on the earth or any other heavenly bodies in space except the space itself. How do know we know that space is absolute inertial system? The law of inertia (Newton's first law of motion), Newton's bucket, and the gyroscopic inertia support the absoluteness of space (see **Chapters 18-19**).

The state of being actually accelerated is qualitatively different from the state of being at rest or in virtual accelerated motion. Therefore, Einstein's relativity of acceleration is wrong. ♦

35

Inertial Mass Is Not the Same as Gravitational Mass

Einstein stated in his chest thought experiment that *inertial mass* is the same as *gravitational mass* (see **Chapter 39**). [1] I do not agree with him. Gravitational mass is often called "weight." *The World Book Encyclopedia* explains *mass* (inertial mass) as follows: [2]

> Mass is often defined as the amount of matter in an object. However, scientists usually prefer to define mass as a measure of *inertia*, a property of all matter. Inertia is the tendency of a stationary object to remain motionless and of a moving object to continue moving at a constant speed and in the same direction.
>
> The greater an object's mass, the more difficult it is to speed it up or slow it down. For example, a railroad locomotive has a greater mass than an automobile. For this reason, it takes more force to stop a moving locomotive than it does an automobile. Force, mass, and acceleration are related by Newton's second law of motion. This law is presented by the equation $F = am$, where F is force, m is mass, and a is acceleration. ...
>
> Mass and weight are not the same thing. Weight is the force on an object due to the pull of earth's gravity. A body weighs less, the farther it gets from the surface of the earth. But its mass remains constant, no matter where it is. For example, crewmen in a spaceship would be "weightless" beyond the earth's gravitational field.

The same encyclopedia explains *weight* as follows: [2'] (the words in the brackets are mine.)

Weight is the gravitational force put forth on an object by the planet on which the object is located. The weight of any object depends on (1) the distance from the object to the center of the planet and (2) the mass (amount of matter) [=inertial mass] of the object.

An object's weight is largest if the object is on the surface of the planet. The weight becomes smaller if the object is moved away for the planet. The object has no weight in space where the gravitational force acting on it is too weak to be measured.

The weight of an object also depends on the mass of the planet. If the mass of a planet is smaller than that of the earth, the gravitational force there is smaller. For example, a man who weighs 200 pounds (91 kilograms0 on the earth would weigh only 32 pounds (15 kilograms) on the moon. He would weigh 76 pounds (34 kilograms0 on Mars, 178 pounds (81 kilograms0 on Venus, and 529 pounds (240 kilograms) on Jupiter.

There is no question about that inertial mass is different from *gravitational mass*: Inertial mass of an object is constant regardless of where the object is. But gravitational mass differs depending on the magnitude of gravitational force acts on the object.

Newton's laws of Motion
(Source: The World Book Encyclopedia, 1979 ed. Article "Motion")

Newton's first law of motion: Any body moving uniformly in a straight line or in a state of rest will remain in uniform motion in a straight line or in state of rest unless it is acted upon by some outside force. The property of matter that tends to keep it in motion or at rest when at rest is called its inertia....

Newton's second law of motion: The change which any force makes in the motion of an object depends upon the size of the force, and mass of the object. The greater the force, the greater the acceleration; the greater the mass of the object, the smaller the acceleration. The motion or the change of motion takes place in the direction in which a force acts.

Newton's third law of motion: For every action there is an equal and opposite reaction. ♦

36

Eight Characteristics That Distinguish Inertial Force from Gravity
(The Principle of Equivalence Is Wrong)

Einstein identified inertial force, which comes from acceleration, with gravity, which comes from the body of matter. Relativists call it the *principle of equivalence* (see **Chapter 39**). [1]

Arthur Beiser states in his textbook that an observer in a closed laboratory cannot distinguish the effects produced by a gravitational field and the inertial effect inside a laboratory being accelerated. [2] But I have found at least eight different characteristics that clearly and readily distinguish inertial effects from gravity. The eight characteristics are, without any order of importance, as follows:

#1. Gravitational force from the body of matter lasts (or seems to last) forever as long as the mass of the body is preserved. But inertial force from acceleration lasts only while an object is being accelerated.

#2. The distance gravitational force reaches is limitless. But the inertial force is limited to the body (laboratory) being accelerated.

#3. The magnitude of gravitational force diminishes as the distance from the center of gravity grows. But the magnitude of inertial force of an accelerated laboratory is the same everywhere inside the laboratory. In the case of rotation motion, the centrifugal force (=inertial force) grows as the distance from the center of the rotation grows.

#4. There is no known means yet to block or adjust the gravitational force from the matter of object. But the direction and magnitude of inertial force of an accelerated body can be readily controlled or adjusted. For example, the direction and size of inertial force inside a spacecraft, for example, are readily adjusted or controlled.

#5. The lines of force of gravity look spread radially from the center of gravity of an object as the spines of a sea urchin spread out from the body of the animal. But the lines of inertial force from acceleration are parallel in the direction of the acceleration.

#6. The acceleration of free-falling caused by natural gravity is uniform. For example, the acceleration of a free-falling body in the earth gravitational field is uniform ($9.8 m/s^2$) (if we ignore the resistance of air). But the artificial acceleration of an elevator or chest, for example, is not always or necessarily uniform. Einstein thought of only the case of uniform acceleration. [20]

#7. Gravity is explained by Newton's law of gravity whereas inertial force is explained by Newton's first law of motion (law of inertia). The law of gravity is clearly different from the law of inertia.

#8. Gravity is one of four ***real forces*** that occur in nature. The real four forces are (1) gravity, (2) electromagnetic force, (3) weak nuclear force, and (4) strong nuclear force. In contrast, inertial force is pseudo (false) force; inertial force is the resistance against the change of speed or direction of motion.

We do not need any more different characteristics to distinguish gravity from inertial force. The nature of gravity from the body of matter is still mystery. We, at least the author, do not fully understand the nature of gravity. But inertial force is fully and clearly explained by Newton's law of motion.

Einstein thought only of uniform inertial force from *uniform acceleration*. But inertial force is not always or necessarily uniform. Inertial force can be uniform, non-uniform, or one-time abrupt one.

Here is a question to relativists: Is the non-uniform inertial force from *non-uniform* acceleration or abrupt one-time jerk (from one-time push/pull) also the same as gravity? The uniform and continuous gravity-like effect is but a *uniform continuous jerk* due to the uniform and continuous change in speed. Continuous and uniform jerk, non-uniform jerk, and one-time jerk are simply inertial effects; these are clearly explained by Newton's first law of motion (the law of inertia).

Some readers might argue that the free-falling motion of an object in the earth's gravitational field is a case of uniform acceleration and that this is the same as the uniform artificial acceleration of a laboratory (chest, spaceship, etc.). They are not the same; there is a fundamental

General Relativity

difference between the two cases: The gravity-like inertial force occurs only in the chest that is being accelerated ***artificially***: There is no gravity-like inertial force inside the free-falling chest in the natural gravitational field. For this reason the author refers to the gravity-like effect in artificially accelerated laboratory as *artificial* or *pseudo* (false) *gravity*." Artificial gravity is fundamentally different from natural gravity.

Einstein's thought process through which he drew the *principle of equivalence* is as follows:

- Gravity causes acceleration.
- The jerk/inertial force that occurs inside an accelerated laboratory (train, chest, etc.) is so similar to the effects cased by gravity that we cannot distinguish one of them from the other.
- Therefore, inertial force is the same as gravitational force.

There are many physical phenomena or effects that look alike superficially yet have different underlying causes. For instance, a train on the railroad can be hauled by men or horses or a steam engine or an electric motor or atomic engine, etc. These various means of hauling the train are the same in the point that they move the railcar. But we cannot conclude that these means are the same in nature or principle. If Einstein were on board the train with his eyes closed and his ears plugged, he might say, "I cannot tell the difference between/among all the hauling means, and I hereby declare a principle that man, horse, steam engine, electric motor, and atomic engine are all the same in nature or principle." Einstein's equivalence principle is wrong.

Einstein was obsessed too much with the idea that he should unify everything into a single equation or principle. So he paid attention mainly to the similar traits of things and overlooked the fundamental differences of physical phenomena. ♦

37

Light Does not Deflect in the Gravitational Field

Beiser explains in his textbook that light deflects in the gravitational field as follows: [1] (The figure numbers are reset for this book.)

> It follows from the principle of equivalence that light should be subject to gravity. If a light beam is directed across an accelerated laboratory, as in **Fig. 37-1**, its path relative to the laboratory will be curved. This means that, if the light beam is subject to the gravitational field to which the laboratory's acceleration is equivalent, the beam would follow the same path.

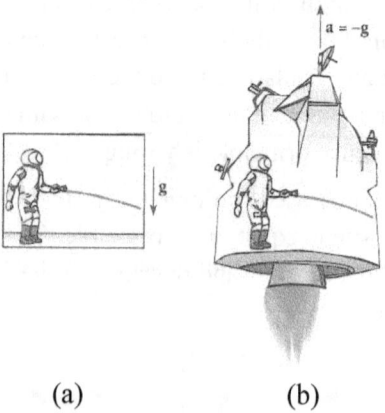

(a) (b)

Fig. 37-1 (a) is a laboratory in gravitational field. (b) is an accelerated Laboratory. According to the principle of equivalence, events that take place in an accelerated laboratory cannot be distinguished from those which take place in a gravitational field. Hence the deflection of a light beam relative to an observer in an accelerated laboratory means that light must be similarly deflected in a gravitational field.

General Relativity

According to general relativity, light rays that graze the sun should have their paths bent toward it by 0.005°—the diameter of a dime seen from a mile away. This prediction was first confirmed in 1919 by photographs of stars that appeared in the sky near the sun during an eclipse, when they could be seen because the sun's disk was covered by the moon. The photographs were then compared with other photographs of the same part of the sky taken then the sun was in a distant part of the sky (see **Fig. 37-2**). Einstein became a world celebrity as a result.

Because light is deflected in a gravitational field, a dense concentration of mass—such as a galaxy of stars—can act as a lens to produce multiple images of a distant light source located behind it.

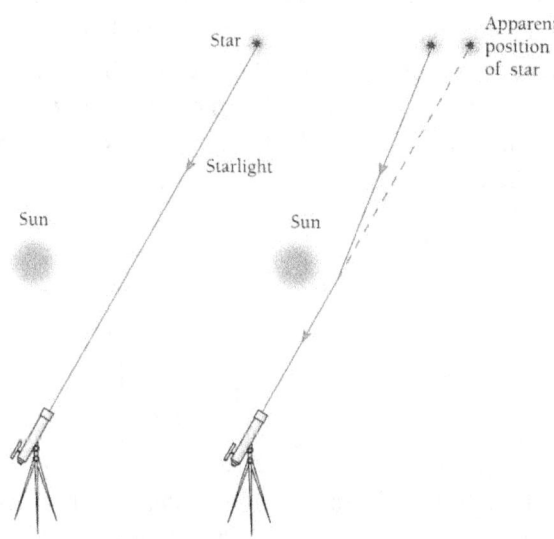

Fig. 37-2 Starlight passing near the sun is deflected by its strong gravitational field. The deflection can be measured during a solar eclipse when the sun's disk is obscured by the moon.

Beiser's explanation, which is based on Einstein's chest thought experiment (see the latter part of this chapter), can be restated as follows:

1. If a laboratory (a chest or spacecraft) is accelerated with respect to inertial system K, and a beam of light, which proceeds in uniform

linear manner in the inertial system, enters the laboratory (through a window on the side wall of the laboratory), the beam of light will deflect when viewed from an observer in the accelerated laboratory.

2. The inertial effects in the accelerated laboratory is the same as the effects of gravity (= the principle of equivalence).

3. Therefore, if starlight passes through the gravitational field of the sun, the starlight should bend (by 0.005°) due to the gravitational field of the sun.

My position is that the speed of light is constant only with respect to absolute space and not with respect to any other coordinate system. Based on my position I will disprove Beiser's (=Einstein's) assertion mentioned above:

Suppose that a chest is accelerated in space (an inertial system) moves at an acceleration of a. A beam of light, which proceeds at uniform linear speed (proper speed) c in space, happens to enter the chest through a window on the side wall of the chest (see **Fig. 37-3**). Then, the light beam will deflect toward the bottom of the chest when viewed by an observer in the chest.

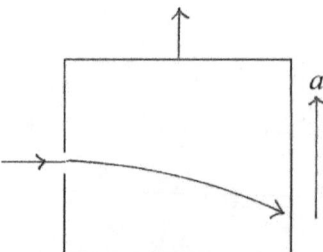

Fig. S-3 A beam of light enters a chest being accelerated. The path of light looks curved when viewed by an observer in the chest.

The deflection of the light beam in curvilinear fashion within the chest is simply *the phenomenon of **Newtonian relativity*** due to the accelerated motion of the chest in the space. For an observer who rests in space (outside the chest), the light beam just proceeds, even after it entered the chest, at the uniform speed (proper speed) of c with respect to absolute space no matter what type of motion the chest is in.

It is true that the light beam curves when viewed from an observer

General Relativity

inside the laboratory being accelerated. For the observer in the laboratory, the curved trajectory of light beam inside the laboratory is obviously longer than the straight path of the light beam viewed from the observer in space (outside the chest). Therefore, the speed (relative speed) of the light is faster than c when viewed from the observer in the laboratory. But for the observer who rests in space, the speed (= the *proper speed*) of light is always c.

In addition, it should be pointed out that the sun's gravitational field diminishes when the distance from the sun grows. I mean the degree of deflection depends on how close or far the starlight passes the surface of the sun. Therefore it is wrong for Beiser (and Einstein) to simply say that the starlight deflects 0.005° when it passes through the sun's gravitational field. In this respect, **Fig. 37-2**, which is from Beiser's textbook, is incorrect. It is not that starlight deflects 0.005 ° regardless of how close or far the starlight passes the sun.

A laboratory needs not be in *accelerated motion* in order to demonstrate the bending of light bean. Light beam will look bent even when the laboratory is in *uniform motion*. In this case, light bean will bend sharply once and then proceeds along the bent path as shown in **Fig. 37-4**. This is also the phenomenon of Newtonian relativity.

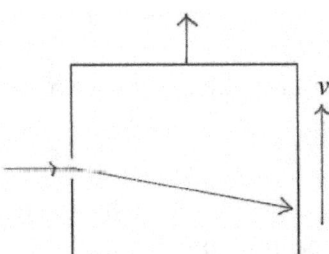

Fig. 37-4 The chest is in uniform motion upward.

In any event, the bent path of light inside the chest, regardless of whether the beam is bent in curvilinear fashion or rectilinear fashion, is longer than the path of light in the inertial frame (space). So the constancy of light speed is disproven. Einstein recognized this in this book *Relativity* by saying as follows (see **Chapter 38**): [2] (The words inside the brackets are mine.)

> In the second place our result [the curving of light beam

> inside the chest in motion] shows that, according to the general theory of relativity, the law of the constancy of the velocity of light in vacuo, which constitutes one of the two fundamental assumptions in the special theory of relativity and to which we have already frequently referred, cannot claim any unlimited validity. A curvature of rays of light can only take place when the velocity of propagation of light varies with position. Now we might think that as a consequence of this, the special theory of relativity and with it the whole theory of relativity would be laid in the dust. But in reality this is not the case. We can only conclude that the special theory of relativity cannot claim an unlimited domain of validity; its results hold only so long as we are able to disregard the influences of gravitational fields on the phenomena (e.g. of light).

Einstein means that special relativity and general relativity are in domain discrimination each other. But Einstein's statement is debunked by his own argument of *the relativity of acceleration* (see **Chapter 34**). [3] According to the relativity of acceleration, there is no qualitative difference between the state of being at rest and the state of being accelerated (this is wrong as we have already seen in **Chapter 34**), and therefore it is contradictory to say that there is domain discrimination between special relativity and general relativity. Actually, in his disc thought experiment, Einstein explained the time dilation of the clock on the rotating disc by both special relativity and general relativity (see **Chapter 40**). [4]

Einstein's prediction that the starlight would bend toward the sun is wrong. Eddington's eclipse observation in 1919 was based on fundamental misconception. The observational evidence (solar eclipse observation done by Eddington) of general relativity is one of the worst manipulations in the history of science (See **Chapter 42**).

What is then gravitational lens effect? I suspect that large bodies of intergalactic gases might be the culprit of the gravitational lens effect. If gravitational lens effect is really due to gravity, such effects should be found everywhere in the sky, especially in the vicinity of large stars or galaxies and not only in some particular stars or galaxies. ♦

General Relativity

38

Is Special Relativity Incompatible with General Relativity?

Einstein acknowledged that the curving of the rays of light relative to an accelerated chest means that the constancy of speed of light is no longer true in general relativity and that special relativity has a limited domain. The following are what Einstein says about the limitation of special relativity in his book *Relativity*: [1] (The term *Galileian domain* or *Galileian body of reference* in the following citation means *free space* without gravitation. The words in the brackets are mine.)

> Let us suppose, for instance, that we know the space-time "course" for any natural process whatsoever, as regards the manner in which it takes place in the Galileian domain relative to a Galileian body of reference K. By means of purely theoretical operation (*i.e.* simply by calculation) we are then able to find how this known natural process appears, as seen from a reference-body K' which is accelerated relative to K. But since a gravitational field exists with respect to this new body of reference K', our consideration also teaches us how the gravitational field influences the process studied.
>
> For example, we learn that a body which is in a state of uniform rectilinear motion with respect to K (in accordance with the law of Galilei) is executing an accelerated and in general curvilinear motion with respect to the accelerated reference-body K' (chest). This acceleration or curvature corresponds to the influence on the moving body of the gravitational field prevailing relativity to K'. It is known that a gravitational field influences the

movement of the bodies in this way, so that our consideration supplies us with nothing essentially new.

However, we obtain a new result of fundamental importance when we carry out the analogous consideration for a ray light. With respect to the Galileian reference-body K, such a ray of light is transmitted rectilinearly with the velocity c, It can be easily be shown that the path of the same ray of light is no longer a straight line when we consider it with reference to the accelerated chest (reference-body K'). From this we conclude, that, in general, rays of light are propagated curvilinearly in gravitational fields. In two respects this result is of great importance.

In the first place, it can be compared with the reality. Although a detailed examination of the question shows that the curvature of light rays required by the general theory of relativity is only exceedingly small for the gravitational fields at our disposal in practice, its estimated magnitude for light rays passing the sun at grazing incidence is nevertheless 1.7 seconds of arc. This ought to manifest itself in the following way. As seen from the earth, certain fixed stars appear to be in the neighborhood of the sun, and are thus capable of observation during a total eclipse of the sun. At such times, these stars ought to appear to be displaced outwards from the sun by an amount indicated above, as compared with their apparent position in the sky when the sun is situated at another part of the heavens. The examination of the correctness or otherwise of this deduction is a problem of the greatest importance, the early solution of which is to he expected of astronomers.[1]

> [Einstein annotated the superscripted part #1 in the same page of his book as follows:
>
> By means of the star photographs of two expeditions equipped by a Joint Committee of the Royal and Royal Astronomical Societies, the existence of the deflection of light demanded by theory was first confirmed during the solar eclipse of 29th May, 1919.]

In the second place our result shows that, according to the general theory of relativity, the law of the constancy of the velocity of light *in vacuo*, which constitutes one of the two fundamental assumptions in the special theory of relativity and to which we have already frequently referred, cannot claim any unlimited validity. A curvature of rays of light can only take place when the velocity of propagation of light varies with position. Now we might think that as a consequence of this, the special

General Relativity

theory of relativity and with it the whole theory of relativity would be laid in the dust. But in reality this is not the case. We can only conclude that the special theory of relativity cannot claim an unlimited domain of validity; its results hold only so long as we are able to disregard the influences of gravitational fields on the phenomena (e.g. of light).

Since it has often been contended by opponents of the theory of relativity that the special theory of relativity is overthrown by the general theory of relativity, it is perhaps advisable to make the facts of the case clearer by means of an appropriate comparison. Before the development of electro-dynamics the laws of electrostatics were looked upon as the laws of electricity. At the present time we know that electric fields can be derived correctly from electrostatic considerations only for the case, which is never strictly realised, in which the electrical masses are quite at rest relatively to each other, and to the co-ordinate system, Should we be justified in saying that for this reason electrostatics is overthrown by the field-equations of Maxwell in electrodynamics? Not in the least. Electrostatics is contained in electro-dynamics as a limiting case: the laws of the latter lead directly to those of the former for the case in which the fields are invariable with regard to time. No fairer destiny could be allotted to any physical theory, than that it should of itself point out the way to the introduction of a more comprehensive theory, in which it lives on as a limiting case.

I am not in the position to mention electrostatics and electro-dynamics because these are not my areas of study. Einstein's statement about these areas, even if it is true, does not prove the verity of relativity because both special relativity and general relativity contain too many fallacies.

According to Einstein, special relativity works only in uniform rectilinear motion in an inertial system and general relativity only in uniformly-accelerated motion. I would indicate that the types of motion of things in reality cannot be clearly classified either as uniform linear motion or uniformly-accelerated motion. Acceleration is not always uniform. In reality, acceleration can be uniform, *accelerated*, irregular, or any combination of uniform motion and non-uniform motion. For example, cars, aircraft, spacecraft, missiles, etc. are normally neither in uniform linear motion nor in uniformly-accelerated motion. Non-uniform acceleration produces non-uniform inertial force. I have a question to relativists: Is non-uniform inertial force from non-uniform acceleration is also the same as gravity? As far as I know, gravity causes uniform acceleration for free-falling bodies.

Einstein usually did his thought experiments in imaginary place called **Galilean space** which he defined as "an imaginary place in the empty space that is far removed from stars and other appreciable masses" (see **Chapter 40**). [4] In reality, our earth is not an inertial system/environment because of its complicated motion consists of rotation, revolution, and wobbling motion (due to the presence of the moon). Then we can say all the experiments or observations performed actually on the earth cannot prove relativity. ♦

General Relativity

39

The Fallacy of Chest Thought Experiment

Einstein explains his chest thought experiment in his book *Relativity* as follows: [1] (The underlines and the serial numbers given to the underlined parts are mine.)

The equality of Inertial and Gravitational Mass as an Argument for the General Postulate of Relativity

We imagine a large portion of empty space, so far removed from stars and other appreciable masses, that we have before us approximately the condition required by the fundamental law of Galilei. It is then possible to choose a Galileian reference-body for this part of space (world), relative to which points at rest remain at rest and points in motion continue permanently in uniform rectilinear motion. As reference-body let us imagine a spacious chest resembling a room with an observer inside who is equipped with apparatus. Gravitation naturally does not exist for this observer. He must fasten himself with string to the floor, otherwise the slightest impact against the floor will cause him to rise slowly towards the ceiling of the room.

To the middle of the lid of the chest is fixed externally a hook with rope attached, and now a "being" (what kind of being is immaterial to us) begins pulling at this with a constant force. The chest together with the observer then begins to move "upwards" with a uniformly accelerated motion. (# 1) In course of time their velocity will reach unheard-of values—provided that we are viewing all this from another reference-body which is not being pulled with a rope.

But how does the man in the chest regard the process? The acceleration of the chest will be transmitted to him by the reaction of the floor of the chest. He must therefore take up this pressure by means of his legs if he does not wish to be laid out full length on the floor. He is then standing in the chest in exactly the same way as anyone stands in a room of a house on our earth. If he releases a body which he previously had in his hand, the acceleration of the chest will no longer be transmitted to this body, and for this reason <u>the body will approach the floor of the chest with an accelerated relative motion</u>. **(# 2)** The observer will further convince himself *that the acceleration of the body towards the floor of the chest is always of the same magnitude, whatever kind of body he may happen to use for the experiment.*

Relying on his knowledge of the gravitational field (as it was discussed in the preceding section), the man in the chest will thus come to the conclusion that he and the chest are in a gravitational field which is constant with regard to time. Of course he will be puzzled for a moment as to why the chest does not fall in this gravitational field. Just then, however, he discovers the hook in the middle of the lid of the chest and the rope which is attached to it, and he consequently comes to the conclusion that the chest is suspended at rest in the gravitational field.

Ought we to smile at the man and say that he errs in his conclusion? I do not believe we ought to if we wish to remain consistent; we must rather admit that his mode of grasping the situation violates neither reason nor known mechanical laws. <u>Even though it is being accelerated with respect to the "Galileian space" first considered, we can nevertheless regard the chest as being at rest. We have thus good grounds for extending the principle of relativity to include bodies of reference which are accelerated with respect to each other, and as a result we have gained a powerful argument for a generalized postulate of relativity</u>. **(# 3)**

We must note carefully that the possibility of this mode of interpretation rests on the fundamental property of the gravitational field of giving all bodies the same acceleration, or, what comes to the same thing, on <u>the law of the equality of inertial and gravitational mass</u>. **(# 4)** If this natural law did not exist, the man in the accelerated chest would not be able to interpret the behavior of the bodies around him on the supposition of a gravitational field, and he would not be justified on the grounds of experience in supposing his reference-body to be "at rest."

General Relativity

Suppose that the man in the chest fixes a rope to the inner side of the lid, and that he attaches a body to the free end of the rope. The result of this will be to stretch the rope so that it will hang "vertically" downwards. If we ask for an opinion of the cause of tension in the rope, the man in the chest will say: "The suspended body experiences a downward force in the gravitational field, and this is neutralized by the tension of the rope; what determines the magnitude of the tension of the rope is the gravitational mass of the suspended body." (# 5) On the other hand, an observer who is poised freely in space will interpret the condition of things thus: "The rope must perforce take part in the accelerated motion of the chest, and it transmits this motion to the body attached to it. The tension of the rope is just large enough to effect the acceleration of the body. That which determines the magnitude of the tension of the rope is the *inertial mass* of the body." (# 6) Guided by this example, we see that our extension of the principle of relativity implies the necessity of the law of the equality of inertial and gravitational mass. Thus we have obtained a physical interpretation of this law. (# 7)

From our consideration of the accelerated chest we see that a general theory of relativity must yield important results on the laws of gravitation. In point of fact, the systematic pursuit of general idea of relativity has supplied the law satisfied by the gravitational field. Before proceeding farther, however, I must warn the reader against a misconception suggested by these considerations. A gravitational field exists for the man in the chest, despite the fact that there was no such field for the co-ordinate system first chosen. Now we might easily suppose that the existence of a gravitational field is always only an *apparent* one. (# 8) We might also think that, regardless of the kind of gravitational field which may be present, we could always choose another reference-body such that no gravitational field exists with reference to it. (# 9) This is by no means true for all gravitational fields, but only for those of quite special form. It is, for instance, impossible to choose a body of reference such that, as judged from it, the gravitational field of the earth (in its entirety) vanishes. (# 10)

We can now appreciate why that argument is not convincing, which we brought forward against the general principle of relativity at the end of Section 18. It is certainly true that the observer in the railway carriage experiences a *jerk* forward as a result of the application of the brake, and that he recognizes in this the non-uniformity of motion (retardation) of the carriage. (# 11) But he is compelled by nobody to refer this jerk to a "real" acceleration (retardation) of the carriage. He might also interpret his experience thus; "My body of reference (the carriage) remains permanently at rest. With reference to it, however, there exists (during the period of the application of the brakes) a gravitational field which is directed forwards and which is variable with respect to time. (# 12) Under the influence of this field, the embankment together with the earth moves nonuniformly in such a manner that their original velocity in the backwards direction is continuously reduced"

We find many fallacies in the thought process of Einstein as follows:

In underlined part # 1, Einstein says that a "being" (what kind of being is immaterial to us) begins pulling at the chest with a constant force so that the chest together with the observer begins to move "upwards" with a uniformly accelerated motion. Einstein says that *what kind of "being"* is immaterial (*unimportant*) for us. But *what kind of being* is a key factor we should note: The *being* must be either God or manmade device such as a space rocket. Since God usually does not participate in physicists' experiment, the being must be man-made device—spacecraft or the like. It is important to make clear that *the being* is not a natural one but an artificial (man-made) one. We will see the importance of this fact in a moment.

In underlined parts # 2 and 3, Einstein identifies the state of being accelerated with the state of being at rest. Relativists might call it *Einstein's relativity of acceleration*. I already proved that the state of being *actually accelerated* is definitely different from the state of being at rest or the state of being *virtually accelerated* (see **Chapter 34**).

In underlined parts # 4~7, Einstein identifies the inertial mass (m) with gravitational mass. And he calls it *the law of the equality of inertial and gravitational mass*. Einstein is wrong. I have already proved in **Chapter 35** that inertial mass is not the same as gravitational mass.

Einstein says, borrowing the mouth of the observer in the chest, "The suspended body experiences a downward force in the gravitational field, and this force is neutralized by the tension of the rope; what determines the magnitude of the tension of the rope is the gravitational mass of the suspended body."

Einstein disregards or forgets Newton's second law of motion in his chest thought experiment. *The World Book Encyclopedia* states in the article "Force" as follows: [2]

> ... Newton's second law of motion gives the relationship between force, mass and acceleration. This law states that every change of motion, or acceleration, is proportional to the force. Physicists express this relationship in the formula $F = ma$, where F is force, m is mass (inertial mass), and a is acceleration. This formula can be rewritten $a = \dfrac{F}{m}$. So, the greater the force, the greater the acceleration will be for the same mass....

Einstein's chest thought experiment is not proper to explain or understand Newton's law of motion. If we are (or Einstein is) to understand Newton's law of motion in Einstein's chest thought experiment, we should be able to change the magnitude of the hauling force (F) of the spacecraft (we have replaced "the being" with a spacecraft), and the mass of the chest (including everything in it). So that, if we change the hauling force (F) of the spacecraft, the acceleration (a) of the chest will be changed proportionally. If we change magnitude of the mass of the chest only (and not the hauling force of the spacecraft), the acceleration (a) of chest will be inversely changed with respect to the size of the mass of the chest.

Einstein did not have the thought of changing the size of the hauling force of the "being" and the total mass of the chest. Einstein did not have the thought of weighing things on the surface of the earth, on the surface of other planet, or in the stationary spacecraft in space (Galilean space. Had Einstein ever done these sorts of experiments, he would not declare the *law of equality of inertial mass and gravitational mass* and *the equality of inertial force and gravity*. Einstein paid attention only to the similarity between the effect of continuous inertial force (*continuous jerk*) and the gravity from the body of matter.

In underlined parts **#8** and **#9**, Einstein identifies artificial acceleration with gravity and then he observes that gravity is "apparent" phenomenon. And he says in underlined part # **10** that gravity is not always apparent by taking the case of earth's gravitational field as an example. Gravity is not an apparent phenomenon. Gravity is actual (real) force.

In underlined parts # **11** and **#12,** Einstein identifies the *jerk*, which passengers in a train experiences when the train is in accelerated/ decelerated motion, with gravity. The "jerk" is an inertial force, not gravity.

Einstein says, in his chest experiment, "It is then possible to choose a Galileian reference-body for this part of space (world), relative to which points at rest remain at rest and points in motion continue permanently in uniform rectilinear motion." This is what Einstein repeated Newton's concept of inertial motion. But Einstein's relativity is not compatible with Newton mechanics. Newton's concept of inertial motion or uniform motion is based upon the *Euclidian geometry* and the absoluteness of space and time. In Einstein physics, the concept of inertial motion (uniform linear motion) is vague and incorrect. Einstein's concept of *linear motion* is not the same as that of Newton. ♦

General Relativity

40

The Fallacy of the Disc Thought Experiment

Einstein identified (misidentified!) centrifugal force (a kind of inertial force) with gravity in his disc thought experiment. Einstein explains his disc thought experiment in his book *Relativity* (fifteenth ed.) as follows: [1] (The underlines and the serial numbers given to the underlined parts are mine. The words in the brackets are mine.)

Let us consider a space-time domain in which no gravitational field exists relative to a reference-body *K* whose state of motion has been suitably chosen. *K* is then a Galilean reference-body as regards the domain considered, and the results of the special theory of relativity hold relative to *K*. Let us suppose the same domain referred to a second body of reference *K'*, which is rotating uniformly with respect to *K*. In order to fix our ideas, we shall imagine *K'* to be in the form of a plane circular disc, which rotates uniformly in its own plane about its centre. An observer who is sitting eccentrically on the disc *K'* is sensible of a force which acts outwards in a radial direction, and which would be interpreted as an effect of inertia (centrifugal force) by an observer who was at rest with respect to the original reference-body *K*. But the observer on the disc may regard his disc as a reference-body which is "at rest"; on the basis of the general principle of relativity he is justified in doing this. (# **1**) The force acting on himself, and in fact on all other bodies which are at rest relative to the disc, he regards as the effect of a gravitational field. (# **2**) Nevertheless, the space-distribution of this gravitational field is of a kind that would not be possible on Newton's theory of gravitation.* (# **3**)

[Einstein annotated the asterisked (*) part above as follows

253

in his book:

> "This field disappears at the center of the disc and increases proportionally to the distance from the center as we proceed outwards."

Einstein means that "gravity (=centrifugal force of the rotating disc)" is zero at the center of the rotating disc, and it (gravity= centrifugal force) grows larger as the distance from the center of the disc grows. Einstein misidentifies centrifugal force with gravity.]

<u>But since the observer believes in the general theory of relativity, this does not disturb him; he is quite in the right when he believes that a general law of gravitation can be formulated—a law which not only explains the motion of the stars correctly, but also the field of force experienced by himself.</u> (# **4**)

The observer performs experiments on his circular disc with clocks and measuring-rods. In doing so, it is his intention to arrive at exact definitions for the signification of time- and space-data with reference to the circular disc K', these definitions being based on his observations. What will be his experience in this enterprise?

<u>To start with, he places one of two identically constructed clocks at the centre of the circular disc, and the other on the edge of the disc, so that they are at rest relative to it.</u> (# **5**) We now ask ourselves whether both clocks go at the same rate from the standpoint of the non-rotating Galilean reference-body K. <u>As judged from this body, the clock at the centre of the disc has no velocity, whereas the clock at the edge of the disc is in motion relative to K in consequence of the rotation.</u> (# **6**) <u>According to the result obtained in Section 12, it follows that the latter clock goes at a rate permanently slower than that of the clock at the centre of the circular disc, i.e. as observed from K. It is obvious that the same effect would be noted by an observer whom we will imagine sitting alongside his clock at the centre of the circular disc.</u> (# **7**) <u>Thus on our circular disc, or, to make the case more general, in every gravitational field, a clock will go more quickly or less quickly, according to the position in which the clock is situated (at rest).</u> (# **8**) For this reason, it is not possible to obtain a reasonable definition of time with the aid of clocks which are arranged at rest with respect to the body of reference. A similar difficulty presents itself when we attempt to apply our earlier definition of simultaneity in such a case, but I do not wish to go any farther into this question.

General Relativity

Moreover, at this stage the definition of the space co-ordinate also presents insurmountable difficulties. <u>If the observer applies his standard measuring-rod (a rod which is short as compared with the radius of the disc) tangentially to the edge of the disc, then, as judged from the Galilean system, the length of this rod will be less than 1, since, according to Section 12, moving bodies suffer a shortening in the direction of the motion.</u> **(# 9)** <u>On the other hand, the measuring-rod will not experience a shortening in length, as judged from K, if it is applied to the disc in the direction of the radius. If, then, the observer first measures the circumference of the disc with his measuring-rod and then the diameter of the disc, on dividing the one by the other, he will not obtain as quotient the familiar number $\pi = 3.14...$, but a larger number,[1]</u> whereas of course, for a disc which is at rest with respect to K, this operation would yield π exactly. **(# 10)**

[Einstein annotated the superscripted part ("a larger number,[1]") as follows:

"Throughout this consideration we have to use the Galilean (non-rotating) system K as reference-body, since we may only assume the validity of the results of the special theory of relativity relative to K (relative to K' a gravitational field prevails)."]

This proves that the propositions of Euclidean geometry cannot hold exactly on the rotating disc, nor in general in a gravitational field, at least if we attribute the length 1 to the rod in all position and in every orientation. Hence the idea of a straight line also loses its meaning. We are therefore not in a position to define exactly the co-ordinates x, y, z relative to the disc by means of the method used in discussing the special theory, and as long as the co-ordinates and times of events have not been defined, we cannot assign an exact meaning to the natural laws in which these occur.

Thus all our previous conclusions based on general relativity would appear to be called in question. In reality we must make a subtle detour in order to be able to apply the postulate of general relativity exactly.

In underlined part # 1 Einstein identifies the state of being at rotation motion (= acceleration motion) with the state of being at rest.

Einstein regards the disc as a coordinate system and he thinks it (disc) can be regarded as being at rest even when it is at rotation motion. Einstein is wrong. Inertial force (centrifugal force) occurs only in the state of being at rotation motion and not in the state of being at rest (see **Chapter 34**).

In underlined part # **2** Einstein identifies centrifugal force (= inertial force) with gravity. Inertial force is not the same as gravity (see **Chapter 36**).

In underlined part # **3** Einstein identifies the centrifugal force with gravity. And he explains that the *spatial-distribution of the gravitational field* on a rotating disc is different from that of Newtonian concept of natural gravitational field. Einstein means that the farther an observer is from the center of a rotating disc, the greater gravitational field (centrifugal force) he would experience. Einstein just misidentifies inertial force (centrifugal force) with gravity.

In underlined part # **4** Einstein again asserts that centrifugal force is the same as gravity.

In underlined part # **5** Einstein identifies the state of being at rotation motion with the state of being at rest.

In underlined part # **6** Einstein assumes that the relative velocity of the clock at the edge of the rotating disc is uniform relative to Galilean frame K. Einstein does not think of the relative speed of the clock (on the disc) with respect to a *specific observer*. If a specific observer is at rest in reference body K, the relative speed of the clock with respect to the observer changes continuously and periodically (confer **Chapter 13**).

In underlined part # **7** Einstein does not think of the relative speed of the clock with respect to a specific observer. Relative speed is defined as the change in the distance between two specific objects.

In underlined part # **8** Einstein thinks that any object in any state of motion can be regarded as being at rest or in inertial state. The state of being at accelerated motion should not be identified with the state of being rest.

In underlined part # **9** Einstein assumes that the velocity (speed + direction) of the clock on the rotating disc is uniform relative to Galilean system K. In this situation, there are no specific observers. The velocity (tangential speed) of the clock on the edge of the rotating disc is a kind of *proper speed* that has nothing to do with observers. Proper speed, which is *observer-free* speed, has nothing to do with relativity.

The relative speed of the clock on the rotating disc with respect to a specific observer, who rests in Galileian space K, varies depending on the position/motion of the clock: For example, the relative speed of the clock at the center of the disc with respect to the observer who rests at the center of the disc is zero (0). However, the relative speed of the clock at the edge of the rotating disc with respect to an observer who rests in Galileian system K changes continuously and periodically (confer **Chapter 13**).

In underlined part # **10**, it is true that the radial distance (the distance from the center of the disc to the clock at the edge of the disc) remains the same. The radius is always perpendicular (90°) to the direction of the tangential motion of the clock at the edge of the rotating disc. If we are to interpret this situation in terms of *cosine effect*, the relative speed of the clock at the edge of the disc with respect to the observer at the center of the disc is always zero (0). In other words, the circumference of the disc and the measuring rod placed tangentially at the edge of the disc have no speed (relative speed zero) with respect to the observer at the center of the disc. Then, the circumference and the rod do not undergo length contraction or time delay even if we assume that the theory of relativity is true. Therefore, if the observer at the center of the disc divides the circumference of the disc by the diameter (two times the radius of the disc), she would obtain the exact π (3.14…).

What if an observer, who rests in Galilean system K, divides the circumference of the rotating disc by the diameter of the disc on the assumption that relativity is true? The relative speed of any matter point or object on the circumference of the disc with respect to the observer is *not* uniform but changes continuously and periodically. Therefore it is wrong to simply say that quotient of the circumference/diameter is larger or smaller than π. ♦

41

Gravity vs. the Geometry of Space-time

Einstein discarded the classical concept of gravity; he replaced gravity with the geometry (curvature) of space-time. Einstein explained that there is no such gravitation of classical concept and that objects simply role down the slope (curvature) of space-time fabric formed as a funnel shape around the larger body of matter.

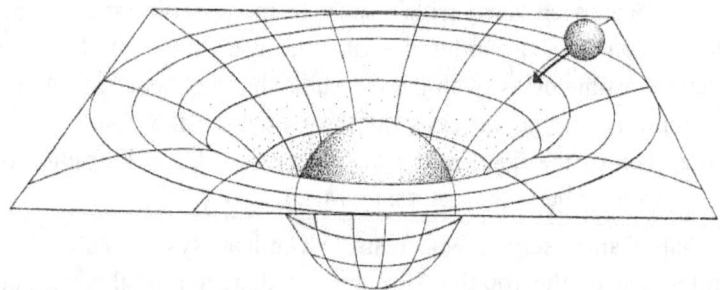

Fig. 41-1 Einsteinian Concept of Gravitation. Objects role down the slope (curvature) of space-time formed around the larger body of matter.

Arthur Beiser illustrates the geometrical concept of gravity in his textbook as follows: [3]

> General relativity pictures gravity as a warping of space-time due to the presence of a body of matter. An object nearby experiences an attractive force as a result of this distortion, much as a marble rolls toward the bottom of a depression in a rubber

General Relativity

sheet. To paraphrase J.A. Wheeler, space-time tells mass how to move, and mass tells space-time how to curve.

There are some problems with the geometrical concept of gravity. The slope (curvature) space-time fabric is formed not only around the larger body of matter; that kind of slope is also formed around the smaller or any body of matter, if Einsteinian concept of gravity is true. Einstein's model is not clear in how the two kinds of slopes interact with each other.

Beiser says that space-time warps due to the presence of a body of matter. Then, what is the agent/cause that makes the space-time warp? In the Newtonian physics, the agent/cause that causes pulling force is called "gravity" whereas relativists simply replace the agent with the curvature of space-time.

But the *slope* or d*epression* of space-time model is not proper to explain the pulling power. Even if a funnel-shape curvature or *slope* of space-time is formed around the body of matter, a marble or iron ball, for example, does not roll down the space-time slope however steep the slope is unless there is pulling force from the body of matter.

People usually think that slope causes any object roll or slide down the slope. This kind of thought is based on the Newtonian concept of *gravity* from the body of matter. If there is no pulling force—whatever we name it---no objet would roll/slide down the slope. Classical physicists call the pulling force *gravity* or *gravitational force*. The space-time geometry model lacks the concept of *energy* or *pulling force*.

Even though we do not fully understand the nature of gravity, we can very successfully explain how it works with Newton's concept of gravity. Newton's concept of gravity is easy to explain and easy to understand. It does not require complicated equations or elusive four-dimension of space-time curvature. The word "space-time tells mass how to move, and mass tells space-time how to curve" sounds like a fairy tale for children. ♦

42

Eddington's Solar Eclipse Observation

Eddington's solar eclipse observation (1919) is known as the proof of general relativity, which is based on the equivalence principle. I have already indicated fundamental problems with general relativity or equivalence principle in the previous chapters. Aside from the controversy over the verity of general relativity itself, Eddington's observation contains many methodological fallacies. The purpose of this chapter is to discuss the methodological fallacies of Eddington's observation and the circumstances Eddington was in.

Fig. 42-1 illustrates Einstein's idea that the starlight passing near the sun should be deflected by the gravitational field of the sun. The idea of Einstein's prediction is as follows: The star in the actual position in the diagram is behind the sun when seen by the observer on the earth. So that the observer on earth cannot normally see the actual star in the present configuration of the heavenly bodies. But the starlight deflects due to the gravity of the sun, and thus the observers on earth can see the star as if it is at the *apparent position*.

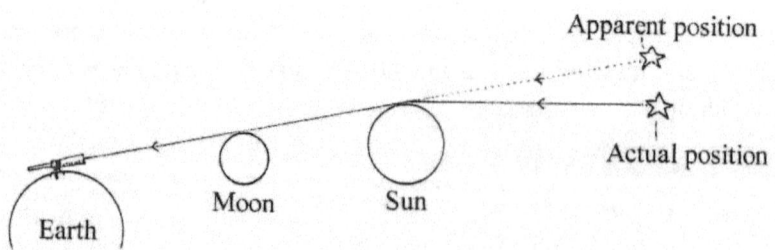

Fig. 42-1 The concept of the deflection of star light.

General Relativity

We need to know a bit about the background of Eddington and the circumstances concerning his solar eclipse observations. Walter Isaacson writes in his Einstein biography as follows: [1]

> Einstein's 1911 paper, "On the Influence of Gravity on the Propagation of Light," and his *Entwurf* equations the following year, had calculated that light would undergo a deflection of approximately (allowing for some data corrections subsequently made) 0.85 arc-second when it passed near the sun, which was the same as would be predicted by an emission theory such as Newton's that treated light as particles. As previously noted, the attempt to test this during the August 1914 eclipse in the Crimea had been aborted by the war, so Einstein was saved the potential embarrassment of being wrong.
>
> Now, according to the field equations he formulated at the end of 1915, which accounted for the curvature of spacetime caused by gravity, he had come up with *twice* that deflection. Light passing next to the sun should be bent, he said, by about 1.7 arc-seconds.
>
> In his 1916 popular book on relativity, Einstein issued yet another call for scientists to test this conclusion. "Stars ought to appear to be displaced outwards from the sun by 1.7 seconds of arc, as compared with their apparent position in the sky when the sun is situated at another part of the heavens," he said. "The examination of the correctness or otherwise of this deduction is a problem of the greatest importance, the early solution of which is to be expected of astronomers."
>
> Willem de Sitter, the Dutch astrophysicist, had managed to send a copy of Einstein's general relativity paper across the English Channel in 1916 in the midst of war and get it to Arthur Eddington, who was the director of the Cambridge Observatory. Einstein was not well-known in England, where scientists then took pride in either ignoring or denigrating their German counterparts. Eddington became an exception. He embraced relativity enthusiastically and wrote an account in English that popularized the theory, at least among scholars.
>
> Eddington consulted with the Astronomer Royal, Sir Frank Dyson, and came up with the audacious idea that a team of English scientists should prove the theory of a German, even as the two nations were at war. In addition, it would help solve a personal problem for Eddington. He was a Quaker and, because of his

pacifist faith, faced imprisonment for refusing military service in England. (In 1918, he was 35 years old, still subject to conscription.) Dyson was able to convince the British Admiralty that Eddington could best serve his nation by leading an expedition to test the theory of relativity during the next full solar eclipse.

That eclipse would occur on May 29, 1919, and Dyson pointed out that it would be a unique opportunity. The sun would then be amid the rich star cluster known as the Hyades, which we ordinary stargazers recognize as the center of the constellation Taurus. But it would not be convenient. The eclipse would be most visible in a path that stretched across the Atlantic near the equator from the coast of Brazil to Equatorial Africa. Nor would it be easy. As the expedition was being considered in 1918, there were German U-boats in the region, and their commanders were more interested in the control of the seas than in the curvature of the cosmos.

Fortunately, the war ended before the expedition began. In early March 1919, Eddington sailed from Liverpool with two teams. One group split off to set up their cameras in the isolated town of Sobral in the Amazon jungle of northern Brazil. The second group, which included Eddington, sailed for the tiny island of Principe, a Portuguese colony a degree north of the equator just off the Atlantic coast of Africa. Eddington set up his equipment on a 500-foot bluff on the island's north tip.

The eclipse was due to begin just after 3: 13 p.m. local time on Principe and last about five minutes. That morning it rained heavily. But as the time of the eclipse approached, the sky started to clear. The heavens insisted on teasing and tantalizing Eddington at the most important minutes of his career, with the remaining clouds cloaking and then revealing the elusive sun.

"I did not see the eclipse, being too busy changing plates, except for one glance to make sure it had begun and another halfway through to see how much cloud there was," Eddington noted in his diary. He took sixteen photographs. "They are all good of the sun, showing a very remarkable prominence; but the cloud has interfered with the star images." In his telegram back to London that day, he was more telegraphic: "Through cloud, hopeful. Eddington."

The team in Brazil had better weather, but the final results had to wait until all the photographic plates from both places could be shipped back to England, developed, measured, and compared.

General Relativity

That took until September, with Europe's scientific cognoscenti waiting eagerly. To some spectators, it took on the postwar political coloration of a contest between the English theory of Newton, predicting about 0.85 arc-second deflection, and the German theory of Einstein, predicting a 1.7 arc-seconds deflection.

The photo finish did not produce an immediately clear result. One set of particularly good pictures taken in Brazil showed a deflection of 1.98 arc-seconds. Another instrument, also at the Brazil location, produced photographs that were a bit blurrier, because heat had affected its mirror; they indicated a 0.86 deflection, but with a higher margin of error. And then there were Eddinton's own plates from Principe. These showed fewer stars, so a series of complex calculations were used to extract some data. They seemed to indicate a deflection of about 1.6 arc-seconds.

Eddington's *voluntary* observation expedition was to kill two birds with one stone for Eddington; Eddington wanted to solve two problems at a time. One was to prove the controversial theory of his personal idol (Einstein) in the enemy nation (Germany). The other was to evade his imprisonment due to his having been avoiding military conscription. He succeeded in solving two problems at one time. But Eddington had to pay price for what he had done; his feat was destined to turn out to be a *historical sham* though not in his time.

Since Eddington had been already an enthusiastic believer of Einstein's theory, he was not a proper person to perform the eclipse observations *objectively* and *impartially*. Eddington's costly and grand-scale observation expedition was to "confirm" Einstein's theory rather than test or check it objectively and impartially. Eddington was the director and supervisor of the entire process of the eclipse observation. He was the *commander-in-chief* in his eclipse observation. There were nobodies who dared to question against the ideas and direction of Eddigton. Including these circumstances, I would indicate several reasons that Eddington's solar eclipse observation was invalid as follows:

#1. It was not proper that Eddington, who had already been an avid believer of Einstein's theory, performed the test of the theory. Psychologically speaking, it is a common sense that such a person usually tries to "confirm" or "assure of" the theory he already believes in rather than test the theory objectively and impartially.

#2. The personal relationship between Eddington and Einstein was unusually close: Both were pacifists and socialists. Due to their common political ideology, which was popular among many intellectuals in Europe at the time of WWI, both Eddington and Einstein were criticized in their own countries respectively. Such a circumstance might have made the two need some sort of solidarity.

#3. The degree of deflection, if any, must vary depending on how close or far the starlight passes the surface of the sun. This factor is important because the magnitude of the sun's gravitational force diminishes as the distance from the surface of the sun grows. This implies that not all the photos taken during the solar eclipse represented the same rate of starlight deflection. If Einstein meant that the rate of starlight deflection was calculated for the moment starlight literally *grazes* the surface of the sun, then such an ideal moment must have not lasted very long during the solar eclipse observation.

The ideal lineup of the star, sun, moon, and earth at the time of the eclipse was ephemeral one. The *quality* of the celestial lineup of the heavenly bodies changes in a matter second or fraction of second. We have the experiences of watching the setting sun sink below the horizon at a considerable speed even with our unaided eye. Eddington must have recorded a series of moments of the *fast-changing relative position* of the sun, earth, and the star. Such photos were taken by old-fashioned manual cameras of the early 20th century technology. Eddington took total sixteen photos during the five-minute duration of the eclipse, roughly at the rate of one photo per 20 seconds. This speed was not fast enough to capture the right moment of the ideal lineup of the heavenly bodies. Since Eddington took a "series" of photos of the eclipse, at least one of them, by chance, could somehow have recorded the close value of the predicted deflection. But that does not necessarily mean that such a photo was taken at the *right moment*. This is similar to the situation that if a student tells a serial numbers of several optional answers to a question, one of the number he told must be the correct answer.

#4. Eddington had known the predicted value of the deflection (1.7 arc-seconds) of the starlight even before he planned the observation expedition. This is somewhat like the situation that a student knows the correct answer before the examination. If an experimenter knows the predicted result before he actually performs the experiment, he is likely to manipulate the outcome (data) or interpret the data to suit the predicted result. Had Eddington not known the predicted value of the deflection (1.7 arc-seconds), he must have ended up with quite different results; he must have not drawn any conclusion.

#5. It seems that Einstein had fiddled with Newton's prediction of star light deflection. In his 1911 paper Einstein predicted that the deflection angle would be 0.85 arc-seconds. [1] This value was the same as the one Newton calculated based on his emission theory. It is not possible that Einstein's calculation (0.85 arc-second) was an innocent coincidence. Einstein's calculation and Newton's calculation were from fundamentally different theories with each other. Einstein must have thought that Newton's prediction was correct. But we still do not know whether Newton's calculation was correct or at least based on correct rationale. In 1915, Einstein corrected his calculation from 0.85 arc-seconds to 1.7 arc-seconds, exactly two times the calculation of Newton. [1] Still Einstein is not immune from our criticism that his calculation was based on Newton's thought. Why exactly two times the value of Newton's prediction?

#6. Eddington (Einstein and Newton also) did not take the thick solar atmospheres into consideration. Since the solar atmospheres are quite thick and always turbulent, they might have influenced the deflection (or *refraction*) of the starlight in *irregular* fashion. The refraction of light and deflection of light are different from each other. *The World Book Encyclopedia* states about the solar atmospheres as follows: [2]

> The sun's surface, or *photosphere* is about 340 miles (547 km) thick, and its temperature is about 10,000° F. (5500° C). The photosphere is actually the innermost layer of the sun's atmosphere. It is from 1 million to 10 million times less dense than water. The photosphere consists of many small patches of gas called *granules*. A typical granule lasts only 5 to 10 minutes, and then it fades away. As old granules fade away, the sun's surface becomes marked with new ones. Scientists believe the granules are caused by waves in the photosphere. The waves are produced by the violent churning motion of the gases in the convection zone....
>
> About 100miles (160 kilometers) above the photosphere, the temperature is about 8000° F. (4400° C). Beyond this point, the temperature rises again. In the *chromosphere* (the middle region of the sun's atmosphere), the temperature may be about 50,000° F.

(27,800° C). The chromosphere consists partly of streams of gas that shoot up briefly. These gas streams, called *spicules*, are about 500 miles (800 kilometers) thick and shoot as high as 10,000 miles (16,000 kilometers). A spicule lasts up to 15 minutes.

Above the chromosphere is a region called the *corona*, which has a temperature of from 2,000,000 to 3,000,000° F. (1,100,000° to 1,670,000° C). The molecules of the corona are so far apart that the gases of the corona have little heat....The temperature drops slowly from the corona outward into space. The corona has no well-defined boundary. Its gases expand constantly away from the sun. This expansion of its gases is called *solar wind*. The temperature of the chromosphere and the corona are a puzzle to astronomers. Heat flows from hot areas to cooler areas, and yet the photosphere is cooler than the two outer regions of the sun's atmosphere. Astronomers believe that the high temperatures of the chromosphere and the corona results from the turbulence of gases in the convection zone....

Prominences are one of the most interesting features of the sun. Each of these bright arches of gas outlines a long, strong bundle of magnetic lines of force. Prominences shine brightly because their gases have a higher density and radiate light more efficiently than do the gases in the chromosphere and corona. A typical prominence may reach 20,000 miles (32,000 kilometers) above the sun's surface. Its total length may be 120,000 miles (190,000 kilometers), and the gases may be 3,000 miles (4,800 kilometers) thick. There are two kinds of prominences—*quiescent* and *active*. A quiescent prominence changes little in appearance during its two- or three-month existence. An active prominence changes rapidly during a period of only several hours. Some active prominences erupt and fling their gases rapidly into space.

The World Book Encyclopedia explains further about the cause of the violent activity on the surface of the sun as follows:

A variety of spectacular activities takes place on the surface of the sun. When these activities become sometimes violent, they are called *solar storms*. Some solar storms occur in the form of huge arches of gas called *prominences*. The arches rise from the edges of the disk and flow back into the sun. Other solar storms occur in the form of areas of gas called sunspots. They appear and disappear in regular cycles. Still other solar storms take place as bright bursts of light flares. Flares release huge amount of solar energy.

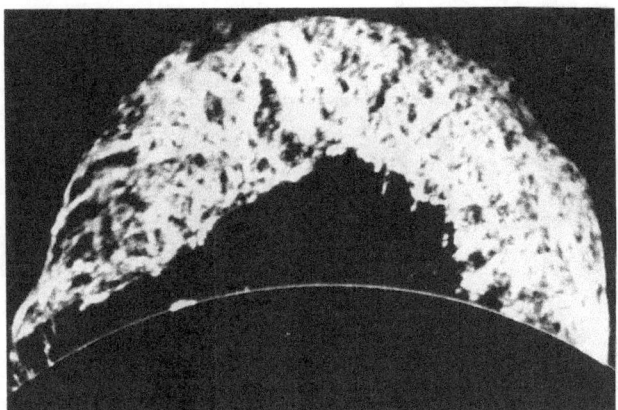

Fig. 42-2 The largest erupting solar prominence that atmosphere have observed occurred on June 4, 1946. Shown in its early stages, it became as large as the sun itself in about one hour.
(Source: The World Book Encyclopedia, 1979 ed.)

Astronomers have found that prominences, sun spots, and other stormy activities on the sun occur because of changes in the patterns of magnetic fields on the sun. A magnetic field occupies the space around a magnet where magnetism exerts a force. Magnetic fields contain magnetic lines of force, or flux lines. In a bar magnet, the lines of force form a simple pattern (...). The sun has a magnetic field that somewhat resembles the pattern of a bar magnet, especially near the sun's poles. But near the sun's equator, the magnetic pattern is always changing because the movement of gases there makes the magnetic field irregular. Atoms of those gases are *ionized*. An ion is an atom of group of atoms that has either gained or lost electrons. Many atoms of gas on the surface of the sun that have lost electrons and form a type of gas called plasma. Particles trapped in a magnetic field usually follow the direction of the magnetic lines of force. But the motion of large quantities of plasma tends to change the direction of these lines. As a result changes occur in the pattern of the sun's magnetic field, and stormy activity takes place.

We marvel at the thickness and violent activities of the solar atmospheres. We also find that the magnetic field on the surface of the sun is not stable. Such violent activities of solar atmospheres are enough to disrupt the idea of *stable rate* of starlight deflection due to the sun's gravitation field.

Fig. 42-3 One of Eddington's photographs of the total solar eclipse of 29 May 1919, presented in his 1920 paper announcing its success, confirming Einstein's theory that light "bends."
(Source: http://en.wikipedia.org/wiki/Arthur_Eddington)

Fig. 42-3 shows the dimension of the corona (during the 1919 solar eclipse observation) in comparison with the size of the sun.

We see the sun on the earth's horizon in the morning before the sun is actually above the horizon because of the ***refraction*** of the sunlight in the earth's atmosphere. (Here let us forget awhile that sunlight takes about 8 minutes to reach the earth.) When the sun is near or on the earth's horizon the shape of the sun looks elliptical. This is due to the refraction of light when it enters a medium of differential density. [3] It is not hard to think that the curve of the starlight during the eclipse must have been due to the *refraction* rather than due to the *deflection* of light.

I once read on the Internet (I cannot fetch the source now) that until the mid-20th century, scientists, including Newton, Galileo, Einstein, Lorentz, etc., did not know about the existence of the sun's atmospheres. Obviously, Einstein and Eddington did not consider the influence of the

General Relativity

solar atmosphere on the curve of starlight.

#7. The principle of Eddington's solar eclipse observation was to compare two kinds of photos—one that shows the position of a star when the sun is not in the scene (so that there is no influence of the sun's gravitational field on the propagation of the starlight) and the other that shows the position of the same star at the *right moment* of the solar eclipse. Concerning this matter, Einstein states in his book *Relativity* (15th ed.) as follows: [4]

> The relative discrepancies to be expected between the stellar photographs obtained during the eclipse and the comparison photographs amounted to a few hundredths of a millimeter only. Thus great accuracy was necessary in making the adjustments required for the taking of the photographs and in their subsequent measurement.

The expected deflection was 1.7 arcs-second. One arc-second is only 1/3600 degree. Yes, great accuracy was called for. But all measurements and comparison were to be done manually because there were no computers or any other advanced technology as we have today.

As far as I know, there are no such records of how or when Eddington ever took *comparison photos*. The comparison photos, if any, should have been identical to the pictures taken during the eclipse except just one factor—the presence of the sun. Theoretically, the comparison photos should not have taken during the eclipse observation. They should have taken much earlier (or later) than the time of eclipse. I mean there should have been considerable amount of time-difference between the two groups of photos. However, time difference means the change of positions of all heavenly bodies involved. So technically, the comparison photos were impossible to take. Even if Eddington ever took the comparison photos, these photos were not valid for the upper mentioned reason.

#8. Granted that Eddington took the two groups of photos—the comparison photos and the eclipse photos—respectively. The size of the image of the star captured in the photos must have been very small but not small enough to be considered a *mathematical point*. A mathematical point has only position and has no size (width, depth, height, etc.) The boundary of the image of the sun on the photos must have not been so sharp and clear-cut. The job of measuring, calculating, and analyzing the value of the deflection of the starlight was done manually by Eddington alone. We may well suspect that such a manual job of a relativity-believer (Eddington) might have incurred quite a degree of human errors (intentional or unintentional).

#9. The output (measurements and analysis) of Eddington were not good to be considered the proof of general relativity. But Eddington pushed his output in the discussion in which Eddington himself was practically the only expert in relativity; and his opinion was chosen in the discussion as the official proof of Einstein's theory. Wikipedia states about the circumstances as follows: [5]

> It has been claimed that Eddington's observations were of poor quality and he had unjustly discounted simultaneous observations at Sobral, Brazil which appeared closer to the Newtonian model. The quality of the 1919 results was indeed poor compared to later observations, but was sufficient to persuade contemporary astronomers. The rejection of the results from the Brazil expedition was due to a defect in the telescopes used which, again, was completely accepted and well-understood by contemporary astronomers. Throughout this period Eddington lectured on relativity, and was particularly well known for his ability to explain the concepts in lay terms as well as scientific. He collected many of these into the *Mathematical Theory of Relativity* in 1923, which Albert Einstein suggested was "the finest presentation of the subject in any language." He was an early advocate of Einstein's General Relativity, and an interesting anecdote well illustrates his humour and personal intellectual investment: Ludwig Silberstein, a physicist who thought of himself as an expert on relativity, approached Eddington at the Royal Society's (6 November) 1919 meeting where he had defended Einstein's Relativity with his Brazil-Principe Solar Eclipse calculations with some degree of scepticism and ruefully charged Arthur as one who claimed to be one of three men who actually understood the theory (Silberstein, of course, was including himself and Einstein as the other two). When Eddington refrained from replying, he insisted Arthur not be "so shy", whereupon Eddington replied, "Oh, no! I was wondering who the third one might be!" ....
>
> Eddington's observations published the next year [1920] confirmed Einstein's theory, and were hailed at the time as

a conclusive proof of general relativity over the Newtonian model. The news was reported in newspapers all over the world as a major story. Afterward, Eddington embarked on a campaign to popularize relativity and the expedition as landmarks both in scientific development and international scientific relations.

It has been claimed that Eddington's observations were of poor quality and he had unjustly discounted simultaneous observations at Sobral, Brazil, which appeared closer to the Newtonian model, but a 1979 re-analysis with modern measuring equipment and contemporary software validated Eddington's results and conclusions. The quality of the 1919 results was indeed poor compared to later observations, but was sufficient to persuade contemporary astronomers. The rejection of the results from the Brazil expedition was due to a defect in the telescopes used which, again, was completely accepted and well-understood by contemporary astronomers.

Throughout this period Eddington lectured on relativity, and was particularly well known for his ability to explain the concepts in lay terms as well as scientific. He collected many of these into the *Mathematical Theory of Relativity* in 1923, which Albert Einstein suggested was "the finest presentation of the subject in any language." He was an early advocate of Einstein's General Relativity, and an interesting anecdote well illustrates his humour and personal intellectual investment: Ludwik Silberstein, a physicist who thought of himself as an expert on relativity, approached Eddington at the Royal Society's (6 November) 1919 meeting where he had defended Einstein's Relativity with his Brazil-Principe Solar Eclipse calculations with some degree of skepticism and ruefully charged Arthur as one who claimed to be one of three men who actually understood the theory (Silberstein, of course, was including himself and Einstein as the other two). When Eddington refrained from replying, he insisted Arthur not be "so shy", whereupon Eddington replied, "Oh, no! I was wondering who the third one might be!"

The Internet website of the Korean Physics Society (KPS) states that the data of the eclipse observation was too much of error and that the final decision was made at a meeting in which Eddington was the only expertise in relativity. [6] The KPS is of and by and for mainstreamers (relativists). The text in the KPS home page seems to be a translation of the original English version. Since I could not find the original English version, I re-translated the Korean text into English as follows: (The words in the brackets are mine.)

Eddington's Solar Eclipse Observation

According to the 1913 gravitation paper of Nordström [not sure of the spelling], special relativity is effective in the gravitational field and the gravitational force is propagated at the same speed of light. The same paper states that the principle of equivalence of Einstein was proven wrong, that light does not deflect even in the strongest gravitational field, and that gravitational force is scalar, not tensor quantity. Nordström's gravitation theory was the counterpart of Einstein's; just whose theory was correct was a great concern for scientists....

Eddington had a close relationship with Frank Watson Dyson who was the most influential astronomer of the British Royal Observatory. Dyson was not a specialist in relativity. However, he became aware of relativity through the influence of Eddington. Dyson was the person who first proposed that English scientists perform the solar eclipse observation to test the theory of Einstein.

Thus the expedition teams were organized, and these teams took the photos of the 1919 solar eclipse to judge whether Einstein's theory was correct. Yet the results of the observation were of too many errors, and these were a problem in approving Einstein's theory. Nevertheless, an urgently organized committee of the British Royal Academic Society and the Royal Astronomical Society issued the conclusion that Einstein's theory had been proven correct. Eddington and Dyson were the most influential figures in drawing the conclusion.

Thus Einstein's theory won against Nordström's, and Einstein was confirmed the hero in the world of science. On November 7, the London Times reported the news with the title "**Scientific Revolution! / New Cosmology! / Newton's Theory Crumbled!**" So Einstein became a celebrity, and the most influential myth of the 20th Century began on November 7, 1919.

Frequently, truth is "elected" by men of power rather than given by nature or God. The theory of relativity was elected the truth mainly on the influence of Eddington. Walter Isaacson writes about the

General Relativity

circumstances in his Einstein's biography as follows: [7]

> The distinguished members of the Royal Society, Britain's most venerable scientific institution, met along with colleges from the Royal Astronomical Society on the afternoon of November 6, 1919, at Burlington House in Piccadilly, for what they knew was likely to be a historic event. There was only one item on the agenda: the report on the eclipse observation.
>
> Sir J, J. Thomson, the Royal Society's president and discoverer of the electron, was in the chair. Alfred North Whitehead, the philosopher, had come down from Cambridge and was in the audience, taking notes. Gazing down on them from an imposing portrait in the great hall was Isaac Newton. "The whole atmosphere of tense interest was exactly that of the Greek drama," Whitehead recorded. "We were the chorus commenting on the decree of destiny … and in the background the picture of Newton to remind us that the greatest of scientific generalization was, now, after more than two centuries, to receive its first modification."
>
> The Astronomer Royal, Sir Frank Dyson, had the honor of presenting the findings. He described in detail the equipment, the photographs, and the complexities of the calculations. His conclusion, however, was simple. "After a careful study of the plates, I am prepared to say that there can be no doubt that they confirm Einstein's prediction," he announced. "The results of the predictions to Sobral and Principe leave little doubt that a deflection of light takes place in the neighborhood of the sun and that it is of the amount demanded by Einstein's generalized theory of relativity."
>
> There was some skepticism in the room. "We owe it to that great man to proceed very carefully in modifying or retouching his law of gravitation," cautioned Ludwig Silverstein, gesturing at Newton's portrait. But it was the commanding giant J. J. Thomson who set the tone. "The result is one of the greatest achievements of human thought," he declared.
>
> Einstein was back in Berlin, so he missed the excitement. He celebrated by buying a new violin. But he understood the historic impact of the announcement that the law of Sir Isaac Newton no longer fully governed all aspects of the universe. "Newton, forgive me," Einstein later wrote, noting the moment. "You found the only way to which, in your age, was just about possible for a man of

highest thought and creative power."

It was a great triumph, but not one easily understood. The skeptical Silberstein came to Eddington and said that people believed that only three scientists in the world understood general relativity. He had been told that Eddington was one of them.

The Quaker said nothing. "Don't be so modest, Eddington!" said Silberstein. Replied Eddington, "On the contrary. I'm wondering who the third might be."

Eddington was not only a prominent astronomer of his day but also an excellent mathematician. This was the factor that helped Eddington become the first disciple of Einstein. Probably he was the only person at his time in the England who could read Einstein's paper written in abstruse math. Sir Frank Dyson was also astronomer. Though Dyson was introduced to the new *way* called relativity by Eddington, he was probably not as bright and enthusiastic in the discipleship as was Eddington.

Eddington and Dyson were the only relativity believers among the distinguished members at the meeting on November 6, 1919; the other members were only ignorant mass when it comes to the *new way* called relativity. Eddington and Dyson were successful in their mission of converting the ignorant mass to a brand new way, and eventually the world accepted the new cult of Einstein. Dyson had convinced the British Admiralty that Eddington could best serve his nation by testing relativity. But Eddington did the opposite; he worst served his great nation. Eddingon not only dishonored Sir Isaacson Newton, the true great hero of England and of the world but also led the world to a fraudulent cult, which no followers truly understand.

There were rumors at that time that only three scientists in the world understood relativity. Two of the three were obviously Einstein and Eddington. But in my opinion, these three persons, who the third might have bee, were those who most misunderstood relativity and misled humanity to non-science. Einstein was a profound deluder and misleader. But it was Eddington's contribution, rather than Einstein's delusion, that ushered in the false revolution of science. Were it not for the contribution of Eddington, Einstein's theory might have died out earlier. Because of the contribution of Eddington, Einstein's teachings (relativity) could have

General Relativity

survived.

Einstein-Eddington relationship reminds me of the relationship of Jesus and Paul. Were it not for the dedication of Paul, Jesus' teachings (Christianity) could not have survived. Marx and Engels were another similar case; were it not for the devotion of Engels, Marx's teachings (communism) might have died out earlier.

Some interesting facts about Eddington's biography read as follows: [8] (Source: http://scienceworld.wolfram.com/biography/Eddington.html)

> Eddington was one of the first to appreciate the importance of Einstein's theories of special and general relativity, and published a treatise on the subject. He led an expedition to observe the total solar eclipse of 1919, in which the bending of light rays predicted by general relativity was observed (although it was later shown that the uncertainties were too large to make any definitive statement).
>
> Eddington was arrogant, and in his later years, cooked up pseudoscientific "proofs" on "physical" grounds that the fine structure constant α was exactly 1/136. When experiments yielded a more accurate value, Eddington produced another proof "proving" that $\alpha \equiv 1/137$. Eddington also disputed Chandrasekhar's use of electron degeneracy pressure to derive the Chandrasekhar mass limit for a white dwarf, insisting that Chandrasekhar failed to understand the difference between "standing" and "progressing" electron waves.
>
> He devoted the last years of his life to writing popular books, and claimed that the number of electrons in the universe is exactly 136×2^{256}, a quantity now known as *the Eddington number*.

Einstein once calculated the diameter of the universe: Einstein found that the diameter of the universe is 320×10^{21} or 320,000,000,000,000,000,000,000 km. [9] What happened to the *Einstein number*? ♦

43

Unified Field Theory

Even before Einstein's time, scientists had been aware of the fact that magnetism and electricity are not entirely different ones but these are related with each other. Maxwell unified magnetism and electricity in 1859. [1] Having been inspired by this achievement, many scientists began to think about the possibility that all natural forces or phenomena could be explained by a single principle. Einstein was one of such scientists. So he unified or identified many different things as the same one. For example,

- Einstein identified proper speed, which is observer-independent, with relative speed, which is observer-dependent.
- Einstein identified the pattern of motion of light (electromagnetic waves) with that of motion of ordinary objects such as train or spacecraft.
- Einstein identified inertial force, which comes from the change in the speed or direction with gravity, which comes from the body of mass.
- Einstein identified the state of being accelerated with the state of being at rest.
- Einstein identified inertial mass with gravitational mass.
- Einstein unified space and time and called it space-time continuum.
- Einstein asserted that the force that governs the orbital motion of planets is the same force that governs the orbital motion of electrons around the nucleus. [2]

General Relativity

- Einstein thought he could unify electromagnetism and gravity.

After the publication of his theory of relativity, Einstein worked further to unify gravity and electromagnetism until the day of his death in vain. There were funny yet thought-provoking anecdotes concerning the unified theory of Einstein. His biography *Einstein, His Life and Universe* reads as follows: [3]

> While others continued to develop quantum mechanics, undaunted by the uncertainties at its core, Einstein persevered in his lonelier quest for a more complete explanation of the universe—a unified field theory that would tie together electricity and magnetism and gravity and quantum mechanics. In the past, his genius had been in finding missing links between different theories. The opening sentences of his 1905 special relativity and light quanta papers were such examples.
>
> He hoped to extend the gravitational field equation of general relativity so that they would describe the electromagnetic field as well. "The mind striving after unification cannot be satisfied that two fields should exist which, by their nature, are quite independent," Einstein explained in his Nobel lecture. "We seek a mathematically unified field theory in which the gravitational field and the electromagnetic field are interpreted only as different components or manifestations of the same uniform field."
>
> Such a unified theory, he hoped, might make quantum mechanics compatible with relativity. He publicly enlisted Planck in this task with a toast at his mentor's sixtieth birthday celebration in 1918: "May he succeed in uniting quantum theory with electrodynamics and mechanics in a single logical system."
>
> Einstein's question was primarily a procession of false steps, marked by increasing mathematical complexity, that began with his reacting to the false steps of others. The first was by the mathematical physicist Hermann Weyl, who in 1918 proposed a way to extend the geometry of general relativity that would, so it seemed, serve as a geometrization of the electromagnetic field as well.
>
> Einstein was initially impressed." It is a first-class stroke of genius," he told Weyl. But he had one problem with it: "I have not been able to settle my measuring-rod objection yet."
>
> Under Weyl's theory, measuring rods and clocks would vary

depending on the path they took through space. But experimental observations showed no such phenomenon. In his next letter, after two more days of reflection, Einstein pricked his bubbles of praise with a wry putdown. "Your chain of reasoning is so wonderfully self-contained," he wrote Weyl. "Except for agreeing with relativity, it is certainly a grand intellectual achievement."

Next came a proposal in 1919 by Theodor Kaluza, a mathematics professor in Königsberg, that a fifth dimension be added to the four dimensions of spacetime. Kaluza further posited that this added spatial dimension was circular, meaning that if you head in its direction you get back to where you started, just like walking around the circumference of a cylinder.

Kaluza did not try to describe the physical reality or location of this added spatial dimension. He was, after all, a mathematician, so he didn't have to. Instead, he devised it as a mathematical device. The metric of Einstein's four-dimensional spacetime required ten quantities to describe all the possible coordinate relationships for any point. Kaluza knew that fifteen such quantities are needed to specify the geometry for a five-dimensional realm.

When he played with the math of this complex construction, Kaluza found that four of the extra five quantities could be used to produce Maxwell's electromagnetic equations. At least mathematically, this might be a way to produce a field theory unifying gravity and electromagnetism.

Once again, Einstein was both impressed and critical. "A five dimensional cylinder world never dawned on me," he wrote Kaluza. "At first glance I like your idea enormously." Unfortunately, there was no reason to believe that most of this math actually had any basis in physical reality. With the luxury of being a pure mathematician, Kaluza admitted this and challenged the physicists to figure it out. "It is still hard to believe that all of these relations in their virtually unsurpassed formal unity should amount to the mere alluring play of a capricious accident," he wrote. "Should more than an empty mathematical formalism be found to reside behind these presumed connections, we would then face a new triumph of Einstein's general relativity."

By then Einstein had become a convert to the faith in mathematical formalism, which had proven so useful in his final push toward general relativity. Once a few issues were sorted out, he helped Kaluza get his paper published in 1921, and followed up later with his own pieces.

The next contribution came from the physicists Oskar Klein, son of Sweden's first rabbi and a student of Niels Bohr. Klein saw a unified field theory not only as a way to unite gravity and electromagnetism, but he also hoped it might explain some of the mysteries lurking in quantum mechanics. Perhaps it could even come up with a way to find "hidden variables" that could eliminate the uncertainty.

Klein was more a physicist than a mathematician, so he focused more than Kaluza had on what the physical reality of a fourth spatial dimension might be. His idea was that it might be coiled up in a circle, too tiny to detect, projecting out into a new dimension from every point in our observable three-dimensional space.

It was all quite ingenious, but it didn't turn out to explain much about the weird but increasingly well-confirmed insights of quantum mechanics or the new advances in particle physics. The Kaluza-Klein theories were put aside, although Einstein over the years would return to some of the concepts. In fact, physicists still do today. Echoes of these ideas, particularly in the form of extra compact dimension, exist in string theory.

Next into the fray came Arthur Eddington, the British astronomer and physicist responsible for the famous eclipse observations. He refined Weyl's math by using a geometric concept known as an affine connection. Einstein read Eddington's ideas while on his way to Japan, and he adopted them as the basis for a new theory of his own. "I believe I have finally understood the connection between electricity and gravitation," he wrote Bohr excitedly. Eddington has come closer to the truth than Weyl."

By now the siren song of a unified theory had come to mesmerize Einstein. "Over it lingers the marble simile of nature," he told Weyl. On his steamer ride through Asia, he polished a new paper and, upon arriving in Egypt in February 1923, immediately mailed it to Planck in Berlin for publication. His goal, he declared, was "to understand the gravitational and electromagnetic field as one."

Once again, Einstein's pronouncements made headlines around the world. "Einstein Describes His Newest Theory," proclaimed the *New York Times*. And once again, the complexity of his approach was played up. As one of the subheads warned:

"Unintelligible to Laymen."

But Einstein told the newspaper it was not all that complicated. "I can tell you in one sentence what it is about," the reporter quoted him as saying. "It concerns the relationship between electricity and gravitation." He also gave credit to Eddington, saying, "It is grounded on the theories of the English astronomer."

In his follow-up articles that year, Einstein made explicit that his goal was not merely unification but finding a way to overcome the uncertainties and probabilities in quantum theory. The title of one 1923 paper stated the quest clearly: "Does the Field Theory Offer Possibilities for the Solution of Quanta Problems?"

The paper began by describing how electromagnetic and gravitational field theories provide causal determinations based on partial differential equations combined with initial conditions. In the realm of the quanta, it may not be possible to choose or apply the initial conditions freely. Can we nevertheless have a causal theory based on field equations?

"Quite certainly," Einstein answered himself optimistically. What was needed, he said, was a method to "overdetermine" the field variables in the appropriate equations. That path of overdetermination became yet another proposed tool that he would employ, to no avail, in fixing what he persisted in calling the "problem" of quantum uncertainty.

Within two years, Einstein had concluded that these approaches were flawed. "My article published [in 1923]," he wrote, "does not reflect the true solution of this problem." But for better or worse, he had come up with yet another method. "After searching ceaselessly in the past two years, I think I have now found the true solution."

His new approach was to find the simplest formal expression he could of the law of gravitation in the absence of any electromagnetic field and then generalize it. Maxwell's theory or electromagnetism, he thought, resulted in a first approximation.

He now was relying more on math than on physics. The metric tensor that he had featured in his general relativity equations had ten independent quantities, but if it were made nonsymmetrical there would be sixteen of them, enough to accommodate electromagnetism.

But his approach led nowhere, just like the others. "The trouble with this idea, as Einstein became painfully aware, is that there really is nothing in it that ties the 6 components of the electric and magnetic fields to the 10 components of the ordinary metric tensor that describes gravitation," says University of Texas physicist Steven Weinberg. "A Lorentz transformation or any other coordinate transformation will convert electric or magnetic fields into mixtures of electric and magnetic fields, but no transformation mixes them with the gravitational field."

Undaunted, Einstein went back to work, this time trying an approach he called "distant parallelism." It permitted vectors in different parts of curved space to be related, and from that sprang new forms of tensors. Most wondrously (so he thought), he was able to come up with equations that did not require that pesky Planck constant representing quanta.

"This looks old-fashioned, and my dear colleagues, and also you, will stick their tongues out because Planck's constant is not in the equation," he wrote Besso in January 1929. "But when they have reached the limit of their mania of the statistical fad, they will return full of repentance to the spacetime picture, and then these equations will form a starting point."

What a wonderful dream! A unified theory without that rambunctious quantum. Statistical approaches turning out to be a passing mania. A return to the field theories of relativity. Tongue-sticking colleagues repenting!

In the world of physics, where quantum mechanics was now accepted, Einstein and his fitful quest for a unified theory were beginning to be seen as quaint. But in the popular imagination, he was still a superstar. The frenzy that surrounded the publication of his January 1929 five-page paper, which was merely the latest in a string of theoretical stabs that missed the mark, was astonishing. Journalists from around the world crowded around his apartment building, and Einstein was barely able to escape them to go into hiding at his doctor's villa on the Havel River outside of town. The *New York Times* had started the drumbeat weeks earlier with an article headlined "Einstein on Verge of Great Discovery: Resents Intrusion."

Einstein's paper was not made public until January 30, 1929, but for the entire preceding month the newspapers printed litany of leaks and speculation. A sampling of the headlines in the *New York Times*, for example, include these:

January 12: Einstein Extends Relativity Theory / New Work

Seeks to Unite Laws of Field of Gravitation and Electro-Magnetism / He Calls It His Greatest 'Book' / Took Berlin Scientist Ten Years to Prepare"

January 19: "Einstein Is Amazed at Stir Over Theory / Holds 100 Journalists at Bay for a Week / BERLIN—For the past week the entire press as represented here has concentrated efforts on procuring the five-page manuscript of Dr. Albert Einstein's 'New Field of Theory.' Furthermore, hundreds of cables from all parts of the world, with prepaid answers and innumerable letters asking for a detailed description or a copy of the manuscript have arrived."

January 25 (page 1): "Einstein Reduces All Physics to One Law / The New Electro-Gravitational Theory Links All Phenomena, Says Berlin Interpreter / Only One Substance Also / Hypothesis Opens Visions of Persons Being Able to Float in Air, Says N.Y.U. Professor / BERLIN—Professor Albert Einstein's newest work, 'A New Field Theory,' which will leave the press soon, reduces to one formula the basic laws of relativistic mechanics and of electricity, according to the person who has interpreted it into English."

Einstein got into the act from his Havel River hideaway. Even before his little paper was published, he gave an interview about it to a British newspaper. "It has been my greatest ambition to resolve the duality of natural laws into unity," he said. "The purpose of my work is to further this simplification, and particularly to reduce to one formula the explanation of the gravitational and electromagnetic fields. For this reason I call it a contribution to 'a unified field theory' ... Now, but only now, we know that the force that moves electrons in their ellipses about the nuclei of atoms is the same force that moves our earth in its annual course around the sun." Of course, it turned out that he did not know that, nor do we know that even now.

He also gave an interview to *Time*, which put him on its cover, the first of five such appearances. The magazine reported that, while the world waited for his "abstruse coherent field theory" to be made public, Einstein was plodding around his country hideaway looking "haggard, nervous, irritate." His sickly demeanor, the magazine explained, was due to stomach ailments and a constant parade of visitors. In addition, it noted, "Dr. Einstein, like so many other Jews and scholars, takes no physical exercise at all."

General Relativity

The Prussian Academy printed a thousand copies of Einstein's paper, an unusually large number. When it was released on January 30, all were promptly sold, and the Academy went back to the printer for three thousand more. One set of pages was pasted in the window of a London department store, where crowds pushed forward to try to comprehend the complex mathematical treatise with its thirty-three arcane equations not tailored for window shoppers. Wesleyan University in Connecticut paid a significant sum for the handwritten manuscript to be deposited as a treasure in its library.

American newspapers were somewhat at a loss. The *New York Herald Tribune* decided to print the entire paper verbatim, but it had trouble figuring out how to cable all the Greek letters and symbols over telegraph machines. So it hired some Columbia physics professors to devise a coding system and then reconstruct the paper in New York, which they did. The *Tribune*'s colorful article about how they transmitted the paper was a lot more comprehensible to most readers than Einstein's paper itself.

The *New York Times*, for its part, raised the unified theory to a religious level by sending reporters that Sunday to churches around the city to report on the sermons about it. "Einstein Viewed as Near Mystic," the headline declared. The Rev. Henry Howard was quoted as saying that Einstein's unified theory supported St. Paul's synthesis and the world's "oneness." A Christian Scientist said it provided scientific backing for Mary Baker Eddy's theory of illusive matter. Others hailed it as "freedom advanced" and a "step to universal freedom."

Theologians and journalists may have been wowed, but physicists were not. Eddington, usually a fan, expressed doubts. Over the next year, Einstein kept refining his theory and insisting to friends that the equations were "beautiful." But he admitted to his dear sister that his work had elicited "the lively mistrust and passionate rejection of my colleagues."

Among those who were dismayed was Wolfgang Pauli. Einstein's new approaches "betrayed" his general theory of relativity, Pauli sharply told him, and relied on mathematical formalism that had no relation to physical realities. He accused Einstein of "having gone over to the pure mathematicians," and he predicted that "within a year, if not before, you will have abandoned that whole distant parallelism, just as earlier you gave up the affine theory."

Pauli was right. Einstein gave up the theory within a year. But he did not give up the quest. Instead, he turned his attention to yet another revised approach that would make more headlines but not more headway in solving the great riddle he had set for himself. "Einstein Completes Unified Field Theory," the New York Times reported on January 23, 1931, with little intimation that it was neither the first nor would it be the last time there would be such announcement. And then again, on October 26 of that year: "Einstein Announces a New Field Theory."

Finally, the following January, he admitted to Pauli, "So you were right after all, you rascal."

And so it went, for another two decades. None of Einstein's offerings ever resulted in successful unified field theory. Indeed, with the discoveries of new particles and forces, physics was becoming less unified. At best, Einstein's effort was justified by the faint praise from the French mathematician Elie Joseph Cartan in 1931: "Even if his attempt does not succeed, it will have forced us to think about the great questions at the foundation of science."

It was true that physics was becoming less unified: Walter Isaacson's *Einstein, His life and Universe* reads as follows: [4]

> In addition, a menagerie of new fundamental particles were discovered beginning in the 1930s. Currently there are dozens of them, ranging from bosons such as photons and gluons to fermions such as electrons, positrons, up quarks, and down quarks. This did not seem to bode well for Einstein's quest to unify everything. His friend Wolfgang Pauli, who joined him at the Institute in 1940, quipped about the futility of this quest. "What has put asunder," he said, "let no man join together."

Human intelligence (animal intelligence as well) is the ability to discern the similarities and differences between/among things or phenomena. Einstein (and his followers) should have been aware of the differences between relative speed and proper speed in the first place. Having been obsessed with the aspiration of unifying everything with an equation, Einstein paid attention mainly to the similarities of things and neglected the differences of things. ♦

APPENDIX-1

Six Exercises of Finding Relative Speed

Exercise A1-1

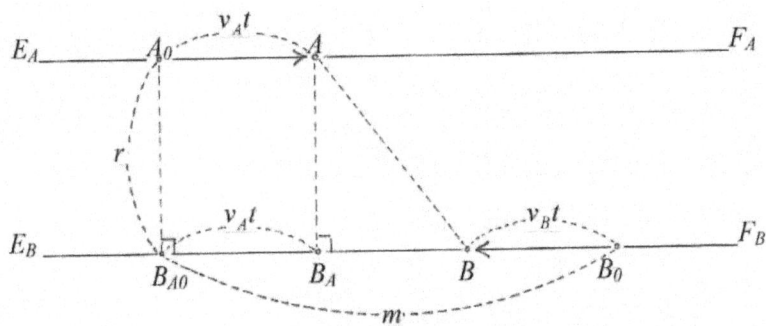

Fig. A1-1

In the above diagram, two straight lines $E_A F_A$ and $E_B F_B$ are parallel with each other. The distance between the two parallel lines is r. Objects A started from point A_0 at time $t = 0$ and moves along $E_A F_A$ in the direction $A_0 \to F_A$ at the speed v_A. That is, $A_0 A = v_A t$. Object B started from point B_0 at time $t = 0$ and moves along straight line $E_B F_B$ in the direction $B_0 \to E_B$ at the speed v_B. That is, $B_0 B = v_B t$. B_{A0} is the point at which the perpendicular from A_0 to $E_B F_B$ meets $E_B F_B$. Since A_0 is a stationary point, B_{A0} is also stationary. $A_0 B_{A0} = r$. B_A is the point at which the perpendicular from A to $E_B F_B$ meets $E_B F_B$. Since A moves at the speed v_A, B_A also moves at the same speed v_A along $E_B F_B$ in the direction $B_{A0} \to F_B$. $B_{A0} B_0 = m$.

Find the following:

(1) The equation that represents distance AB with respect to time t.

(2) The equation that represents the speed of A relative to B.

(3) The speed of A relative to B at the moment AB becomes perpendicular to both $E_A F_A$ and $E_B F_B$.

Solution:

(1) Let distance AB be Y. From right triangle $A B_A B$,

$$Y = AB = \sqrt{(B_A B)^2 + (AB_A)^2}$$

(Since $B_A B = |m - (v_B t + v_A t)|$ and $AB_A = r$,)

$$= \sqrt{\{(v_B t + v_A t) - m\}^2 + r^2} \ . \qquad \text{(A1-1)}$$

Eq. (A1-1) represents distance AB at time t.

(2) The derivative of Y with respect to time t is

$$Y' = \frac{dy}{dt} = \frac{(v_A + v_B)[(v_A + v_B)t - m]}{\sqrt{[(v_A + v_B)t - m]^2 + r^2}} . \qquad \text{(A1-2)}$$

Eq. (A1-2) represents the relative speed of A with respect to B.

How do we know whether **Eq. (A1-2)** is correct?
We can prove **Eq. (A1-2)** by checking the following cases:
If $v_A \neq 0$ and $v_B = 0$,

$$Y' = \frac{(v_A + v_B)[(v_A + v_B)t - m]}{\sqrt{[(v_A + v_B)t - m]^2 + r^2}} = \frac{(v_A)(v_A t - m)}{\sqrt{(v_A t - m)^2 + r^2}} .$$

[This is the same form as **Eq. (2)** in **Chapter 4**.]

If $v_A = 0$ and $v_B \neq 0$,

Appendix-1

$$Y' = \frac{(v_A+v_B)[(v_A+v_B)t-m]}{\sqrt{[(v_A+v_B)t-m]^2+r^2}} = \frac{v_B(v_Bt-m)}{\sqrt{(v_Bt-m)^2+r^2}}.$$

[This is the same form as **Eq. (2)**.]

If $v_A \neq 0$, $v_B \neq 0$, and $m = 0$,

$$Y' = \frac{(v_A+v_B)[(v_A+v_B)t-m]}{\sqrt{[(v_A+v_B)t-m]^2+r^2}} = \frac{v_B(v_Bt)}{\sqrt{(v_Bt)^2+r^2}}.$$

[This is the same form as **Eq. (3)** in **Chapter 4**.]

If $v_A \neq 0$, $v_B \neq 0$, $m \neq 0$, and $r = 0$,

$$Y' = \frac{(v_A+v_B)[(v_A+v_B)t-m]}{\sqrt{[(v_A+v_B)t-m]^2+r^2}} = \frac{(v_A+v_B)[(v_A+v_B)t-m]}{\sqrt{[(v_A+v_B)t-m]^2}}$$

$$= v_A+v_B.$$

(This is the case in which A and B move on the same straight line in the opposite directions of each other. This is a 1-D situation.)

If $v_A \neq 0$, $v_B = 0$, and $r = 0$,

$$Y' = \frac{(v_A+v_B)[(v_A+v_B)t-m]}{\sqrt{[(v_A+v_B)t-m]^2+r^2}} = \frac{v_A(v_At-m)}{\sqrt{(v_At-m)^2}} = v_A.$$

(This is a 1-D situation: B rests on the line of motion of A.)

If $v_A = 0$, $v_B \neq 0$, and $r = 0$,

$$Y' = \frac{(v_A+v_B)[(v_A+v_B)t-m]}{\sqrt{[(v_A+v_B)t-m]^2+r^2}} = \frac{v_B(v_Bt-m)}{\sqrt{(v_Bt-m)^2}} = v_B.$$

(This is a 1-D situation: A rests on the line of motion of B.)

The above results prove that **Eq. (A1-2)** is correct.

(3) From $v_A t = m - v_B t$, the time AB becomes perpendicular to both $E_A F_A$ and $E_B F_B$ is $t = \dfrac{m}{v_A+v_B}$.

If $t = \dfrac{m}{v_A + v_B}$ in **Eq. (A1-2)**,

$$Y' = \frac{(v_A + v_B)[(v_A + v_B)t - m]}{\sqrt{[(v_A + v_B)t - m]^2 + r^2}} = \frac{(v_A + v_B)(m - m)}{\sqrt{(m - m)^2 + r^2}} = 0.$$

(This is when the observation angle of A with respect to B is 90°.) This result confirms *the third law of relative speed* (see **Chapter 4**). ♦

Exercise A1-2

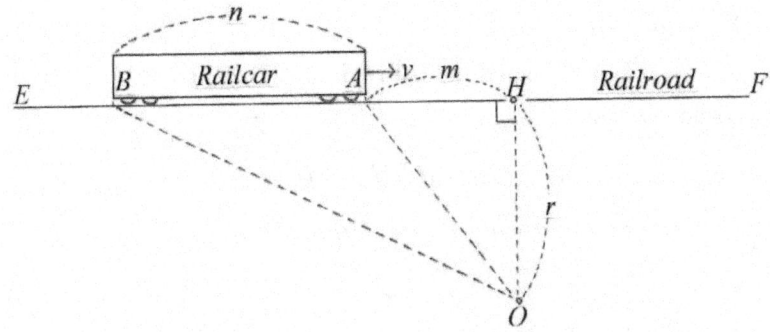

Fig. A1-2 Conductor A is at the front part of the railcar, and conductor B is at the rear part of the rail car. Observer O is distance r from the railroad.

In the diagram above, a railcar moves on a straight railroad at the speed of v in the direction $E \rightarrow F$. Conductor A is at the front end of the railcar, and conductor B is at the tail end of the railcar. The distance between the two conductors is n ($AB = n$). Observer O is distance r from railroad EF. H is the point at which the perpendicular from O to EF meets EF. $OH = r$. The diagram shows the moment A is distance m from H ($AH = m$). Suppose that the time (t) at the moment conductor A is distance m from H is zero ($t = 0$).

Find the following:

(1) The equation that represents distance OA and the equation that represents distance OB.
(2) The equation that represents the speed of A with respect to O and the equation that represents the speed of B with respect to O.
(3) The speed of A with respect to O and the speed of B with respect to O at the moment A passes point H.
(4) The speed of A with respect to O and the speed of B with respect to O at the moment B passes point H.

Solution:

(1) Let distance OA be Y_A.

From right triangle AOH, $OA = \sqrt{(AH)^2 + (OH)^2}$.

(Since $AH = |m - vt|$, and $OH = r$,)

$$Y_A = OA = \sqrt{(vt - m)^2 + r^2}. \tag{A1-3}$$

[**Eq. (A1-3)** is the equation that represents distance AO.]
Let distance OB be Y_B.
From right triangle BOH, $OB = \sqrt{(BH)^2 + (OH)^2}$.

(Since $BH = |m + n - vt|$ and $OH = r$,)

$$Y_B = (OB) = \sqrt{(vt - m - n)^2 + r^2}. \tag{A1-4}$$

[**Eq. (A1-4)** is the equation that represents distance BO.]
(2) The derivative of Y_A [**Eq. (A1-3)**] with respect to time t is

$$Y'_A = \frac{v(vt - m)}{\sqrt{(vt - m)^2 + r^2}}. \tag{A1-5}$$

[**Eq. (A1-5)** is the equation that represents the relative speed of A with respect to O.)
The derivative of Y_B [**Eq. (A1-4)**] with respect to time t is

$$Y'_B = \frac{v(vt - m - n)}{\sqrt{(vt - m - n)^2 + r^2}}. \tag{A1-6}$$

[**Eq. (A1-6)** represents the relative speed of B with respect to O.]
(3) The time conductor A passes point H is $t = \dfrac{m}{v}$.

From **Eq. (A1-5)**, the speed of A with respect to O at $t = \dfrac{m}{v}$ is

$$Y'_A = \frac{v(vt - m)}{\sqrt{(vt - m)^2 + r^2}} = \frac{v(m - m)}{\sqrt{(m - m)^2 + r^2}} = 0.$$

From **Eq. (A1-6)**, the speed of B relative to O at $t = \dfrac{m}{v}$ is

$$Y'_B = \frac{v(m - m - n)}{\sqrt{(m - m - n)^2 + r^2}} = \frac{-nv}{\sqrt{n^2 + r^2}}.$$

The negative value of Y'_B ($= \dfrac{-nv}{\sqrt{n^2 + r^2}}$) means that the distance BO decreases at the moment $t = \dfrac{m}{v}$.

Appendix-1

We can see that the speed of A relative to O at $t = \dfrac{m}{v}$ is not equal to that of B with respect to O at $t = \dfrac{m}{v}$.

(4) The time conductor B passes point H is $t = \dfrac{m+n}{v}$.

From **Eq. (A1-6)**, the speed of B relative to O at $t = \dfrac{m+n}{v}$ is

$$Y'_B = \dfrac{v(m+n-m-n)}{\sqrt{(m+n-m-n)^2 + r^2}} = 0.$$

From **Eq. (A1-5)**, the speed of A with respect to O at $t = \dfrac{m+n}{v}$ is

$$Y'_A = \dfrac{v(m+n-n)}{\sqrt{(m+n-n)^2 + r^2}} = \dfrac{vm}{\sqrt{m^2 + r^2}}.$$

The value of $\dfrac{vm}{\sqrt{m^2 + r^2}}$ is positive. This means that distance OA increases at the moment $t = \dfrac{m+n}{v}$.

Conclusion: The relative speed of the *front part* of the railcar with respect to an observer who is off the line of motion of the railcar is different from the relative speed of the *tail part* of the railcar with respect to the observer at any moment.

If we assume that the railcar is a segment of line and every point on the line is a conductor, the speeds of all the conductors with respect to observer O who is off the line of motion of the railcar are all different. ♦

Exercise A1-3

(This exercise is the same as the case we dealt in **Chapter 9**. This exercise is a prerequisite for Exercises **A1-4**, **A1-5**, and **A1-6** which follow.)

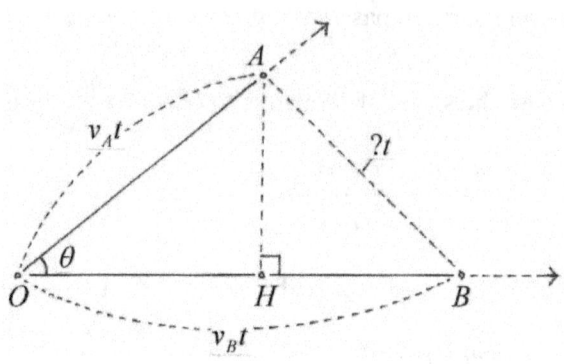

Fig. A1-3

In the above diagram, two objects A and B started from common origin O at the same time and move at the uniform speeds of v_A and v_B respectively maintaining intersection angle $AOB = \theta$. H is the point at which the perpendicular from A to OB meets OB. Since A moves, H moves also accordingly.

Find the relative speed of A with respect to B.

Solution:

The given situation is not a 1-D situation because A and B are not situated/moving on the same straight line. If the intersection angle AOB ($=\theta$) is either 0° or 180°, it is a 1-D situation.

If $\theta = 0°$ (this is when A and B move in the same direction), the relative speed of A with respect to B is $v_A \sim v_B$.

If $\theta = 180°$ (this is when two objects move in the opposite direction), the relative speed of A with respect to B is $v_A + v_B$.

If $0° < \theta < 180°$ (this is a 2-D situation), we should use calculus and trigonometry as follows:

Let distance AB be Y.

Appendix-1

Since AHB is a right triangle,

$$Y = AB = \sqrt{(BH)^2 + (AH)^2}$$

(Since $BH = OB - OH = v_B t - v_A t \cos\theta$, and $AH = v_A t \sin\theta$,)

$$= \sqrt{(v_B t - v_A t \cos\theta)^2 + (v_A t \sin\theta)^2}$$

$$= t\sqrt{(v_B)^2 + (v_A)^2 (\sin^2\theta + \cos^2\theta) - 2v_A v_B \cos\theta}$$

(since $\sin^2\theta + \cos^2\theta = 1$,)

$$= t\sqrt{(v_A)^2 + (v_B)^2 - 2v_A v_B \cos\theta}. \qquad (A1\text{-}7)$$

Eq. (A1-7) represents distance AB at time t.

The derivative of **Eq. (A1-7)** with respect to time t is

$$Y' = \sqrt{(v_A)^2 + (v_B)^2 - 2v_A v_B \cos\theta} \qquad (A1\text{-}8)$$

Eq. (A1-8) represents the relative speed of A with respect to B. Note that there is no time factor t in **Eq. (A1-8)**. This means that the relative speed of A with respect to B is constant or independent of time. **Fig. (A1-3)** is the case in which angle $ABO < 90°$. We find the same results [**Eqs. (A1-7, A1-8)**] when angle $ABO \geq 90°$.

How do we know whether **Eq. (A1-8)** is correct? We can prove **Eq. (A1-8)** by replacing θ with $0°$, $90°$, or $180°$; or by replacing either v_A or v_B with zero (0).

If $\theta = 0°$, $v_A \neq 0$, and $v_B \neq 0$,

$$Y' = \sqrt{(v_A)^2 + (v_B)^2 - 2v_A v_B \cos 0°} \quad \text{(Since } \cos 0° = 1,\text{)}$$

$$= \sqrt{(v_A)^2 + (v_B)^2 - 2v_A v_B} = |v_A - v_B| \text{ or } v_A \sim v_B.$$

(This is a 1-D situation: A and B move in the same straight line in the same direction.)

If $\theta = 180°$, $v_A \neq 0$, and $v_B \neq 0$,

$$Y' = \sqrt{(v_A)^2 + (v_B)^2 - 2v_A v_B \cos 180°}$$

(since $\cos 180° = -1$,)

$$= \sqrt{(v_A)^2 + (v_B)^2 + 2v_Bv_B} = v_A + v_B.$$

(This is 1-D situation.)

If $\theta = 90°$, $v_A \neq 0$, and $v_B \neq 0$,

$$Y' = \sqrt{(v_A)^2 + (v_B)^2 - 2v_Av_B \cos 90°}$$

(Since $\cos 90° = 0$,)

$$= \sqrt{(v_A)^2 + (v_B)^2}.$$

(Note that *AOB* is a right triangle.)

If $v_A = 0$ and $v_B \neq 0$,

$$Y' = \sqrt{(v_A)^2 + (v_B)^2 - 2v_Av_B \cos \theta} = v_B.$$

(This is a 1-D situation: A stays at point O.)

If $v_B = 0$ and $v_A \neq 0$,

$$Y' = \sqrt{(v_A)^2 + (v_B)^2 - 2v_Av_B \cos \theta} = v_A.$$

(This is a 1-D situation: B stays at point O.)

Therefore, we can see that **Eq. (A1-8)** is correct.

Eq. (A1-8) is good for any value of v_A, v_B, or θ. For example, if $v_A = 40$ km/sec, $v_B = 70$km/sec, and $\theta = 24.5°$, the relative speed of A with respect to B is

$$\sqrt{(40km/\sec)^2 + (70km/\sec)^2 - 2(40km/\sec)(70Km/sex)\cos 24.5°}$$
$$\approx 37.47\text{km/sec. Note: } \cos 24.5° = 0.9100 \text{ km/sec.} \blacklozenge$$

Exercise A1-4

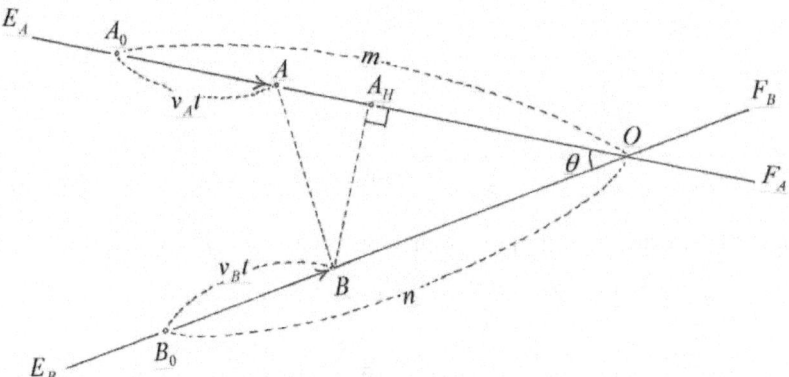

Fig. A1-4

In the diagram above, object A started from point A_0 at time $t = 0$ and moves along straight line $E_A F_A$ in the direction $E_A \to F_A$ at the speed v_A. That is, $A_0 A = v_A t$. Object B started from point B_0 at time $t = 0$ and moves along straight line $E_B F_B$ in the direction $E_B \to F_B$ at the speed v_B. That is, $B_0 B = v_B t$. Two straight lines $E_A F_A$ and $E_B F_B$ intersect at point O. A_0 is distance m from O. B_0 is distance n from O. Intersection angle $A_0 O B_0 = \theta$. A_H is the point at which the perpendicular from B to $E_A F_A$ meets $F_A F_A$. Since B moves, A_H also moves accordingly along $E_A F_A$.

Find the following:

(1) The equation that represents distance AB.

(2) The equation that represents the speed of A relative to B.

Solution:

(1) Let distance AB be Y. Since $A A_H B$ is a right triangle,

$$AB = Y = \sqrt{(AA_H)^2 + (BA_H)^2}$$

(since $A A_H = A_0 O - A_0 A - A_H O = m - v_A t - BO\cos\theta$, and $B A_H = BO\sin\theta$,)

295

$$= \sqrt{(m - v_A t - BO\cos\theta)^2 + (BO\sin\theta)^2}$$

(since $BO = n - v_B t$,)

$$= \sqrt{(m - v_A t)^2 + (n - v_B t)^2 (\cos^2\theta + \sin^2\theta) - 2(m - v_A t)(n - v_B t)\cos\theta}$$

$$= \sqrt{(m - v_A t)^2 + (n - v_B t)^2 - 2(m - v_A t)(n - v_B t)\cos\theta}. \quad \text{(A1-9)}$$

Eq. (A1-9) is the equation that represents distance AB.

(2) The derivative of **Eq. (A-9)** with respect to time t is

$$Y' = \frac{(v_A t - m)(v_A - v_B \cos\theta) + (v_B t - n)(v_B - v_A \cos\theta)}{\sqrt{(m - v_A t)^2 + (n - v_B t)^2 - 2(m - v_A t)(n - v_B t)\cos\theta}}. \quad \text{(A1-10)}$$

Eq. (A1-10) is the equation that represents the speed of A with respect to B.

How do we know whether **Eq. (A1-10)** is correct?

We can prove **Eq. (A1-10)** by replacing both m and n with zero (0) or by replacing θ with 0° or 180°.

If $m = 0$, and $n = 0$ in **Eq. (A1-10)**,

$$Y' = \frac{(v_A t - m)(v_A - v_B \cos\theta) + (v_B t - n)(v_B - v_A \cos\theta)}{\sqrt{(m - v_A t)^2 + (n - v_B t)^2 - 2(m - v_A t)(n - v_B t)\cos\theta}}$$

$$= \frac{t\{(v_A)^2 + (v_B)^2 - 2v_A v_B \cos\theta\}}{t\sqrt{(v_B)^2 + (v_B)^2 - 2v_A v_B \cos\theta}}$$

$$= \sqrt{(v_A)^2 + (v_B)^2 - 2v_A v_B \cos\theta}. \text{ [This is the same as } \textbf{Eq. (A1-8)}.]$$

If θ is 0° in **Eq. (A1-10)**,

$$Y' = \frac{(v_A t - m)(v_A - v_B \cos\theta) + (v_B t - n)(v_B - v_A \cos\theta)}{\sqrt{(m - v_A t)^2 + (n - v_B t)^2 - 2(m - v_A t)(n - v_B t)\cos\theta}}$$

(since $\cos 0° = 1$,)

$$= \frac{(v_A t - m)(v_A - v_B) + (v_B t - n)(v_B - v_A)}{\sqrt{(m - v_A t)^2 + (n - v_B t)^2 - 2(m - v_A t)(n - v_B t)}}$$

Appendix-1

$$= \frac{(v_A t - m)(v_A - v_B) + (v_B t - n)(v_B - v_A)}{\sqrt{[(m - v_A t) - (n - v_B t)]^2}}$$

$$= \frac{(m - v_A t)(v_B - v_A) + (v_B t - n)(v_B - v_A)}{(m - v_A t) - (n - v_B t)}$$

$$= \frac{(v_B - v_A)[(m - v_A t) - (n - v_B t)]}{(m - v_A t) - (n - v_B t)} =$$

$$= v_B - v_A \text{ or } v_B \sim v_A.$$

(This is 1-D situation; object A and object B move in the same direction on the same straight line.)

If θ is 180° in **Eq. (A1-10)**,

$$Y' = \frac{(v_A t - m)(v_A - v_B \cos\theta) + (v_B t - n)(v_B - v_A \cos\theta)}{\sqrt{(m - v_A t)^2 + (n - v_B t)^2 - 2(m - v_A t)(n - v_B t)\cos\theta}}$$

(since $\cos 180° = -1$,)

$$= \frac{(v_A t - m)(v_A + v_B) + (v_B t - n)(v_B + v_A)}{\sqrt{(m - v_A t)^2 + (n - v_B t)^2 + 2(m - v_A t)(n - v_B t)}}$$

$$= \frac{(v_A t - m)(v_A + v_B) + (v_B t - n)(v_B + v_A)}{\sqrt{[(m - v_A t) + (n - v_B t)]^2}}$$

$$= \frac{(v_B + v_A)[(m - v_A t) + (n - v_B t)]}{(m - v_A t) + (n - v_B t)}$$

$$= v_B + v_A.$$

(This is 1-D situation; object A and object B move in the opposite direction of each other on the same straight line.)

Therefore, we can see that **Eq. (A1-10)** is correct. ◆

Exercise A1-5

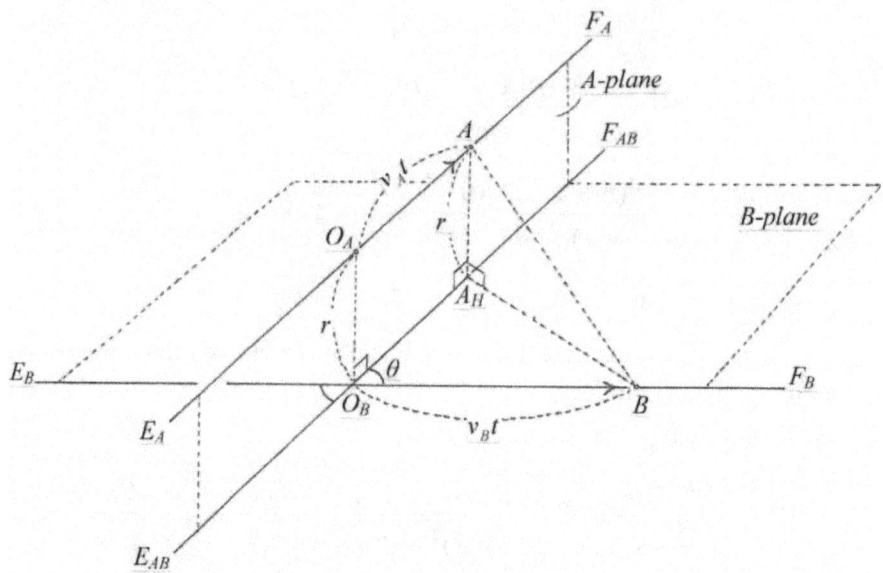

Fig. A1-5

In the above diagram, two straight lines $E_A F_A$ and $E_B F_B$ are not in the same plane. Point O_A is on straight line $E_A F_A$, and point O_B is on straight line $E_B F_B$. $O_A O_B$ is the shortest distance that connects the two straight lines $E_A F_A$ and $E_B F_B$. Straight line $E_{AB} F_{AB}$, which contains O_B, is parallel to $E_A F_A$. Now, $E_{AB} F_{AB}$ is the orthogonal projection of $E_A F_A$ on B-plane, which contains two straight lines $E_B F_B$ and $E_{AB} F_{AB}$. A-plane, which contains two parallel lines $E_A F_A$ and $E_B F_B$, is perpendicular to B-plane. Intersection angle $F_{AB} O_B F_B = \theta$. Object A started from O_A at time $t = 0$ and moves along straight line $E_A F_A$ in the direction $E_A \to F_A$ at the speed v_A. That is, $O_A A = v_A t$. Object B started from O_B at time $t = 0$ and moves along straight line $E_B F_B$ in the direction $E_B \to F_B$ at the speed v_B. That is, $O_B B = v_B t$. A_H is the point at which the perpendicular from A to $E_{AB} F_{AB}$ meets $E_{AB} F_{AB}$. Since A moves, A_H moves accordingly along $E_{AB} F_{AB}$ at the speed v_A. $O_A O_B = A A_H = r$.

Appendix-1

Find the following:

(1) The equation that represents distance AB with respect to time t.

(2) The equation that represents the speed of A relative to B.

(3) The relative speed between A and B at time $t = 0$.

Solution:

(1) Let distance AB be Y. Since $E_{AB}F_{AB}$ is the orthogonal projection of $E_A F_A$ on B-plane, angle $AA_H B$ is right angle. From right triangle $AA_H B$,

$$AB = Y = \sqrt{(AA_H)^2 + (A_H B)^2}.$$

{Since $AA_H = r$, and $A_H B = t\sqrt{(v_A)^2 + (v_B)^2 - 2v_A v_B \cos\theta}$ (see **Eq. (A1-7)** of **Exercise A1-3**),}

$$Y = \sqrt{(v_A t)^2 + (v_B t)^2 - 2v_A v_B t^2 \cos\theta + r^2}. \qquad \text{(A1-11)}$$

Eq. (A-11) represents distance AB at time t.

(2) The derivative of Y with respect to time t is

$$Y' = \frac{(v_A)^2 t + (v_B)^2 t - 2v_A v_B t \cos\theta}{\sqrt{(v_A t)^2 + (v_B t)^2 - 2v_A v_B t^2 \cos\theta + r^2}}. \qquad \text{(A1-12)}$$

Eq. (A1-12) is the relative speed equation of A with respect to B.

How do we know whether **Eq. (A1-12)** is correct?

We can prove **Eq. (A1-12)** by replacing either v_A or v_B with zero (0).

If $v_A = 0$ in **Eq. (A1-12)** (This is when A stays at point O_A),

$$Y' = \frac{(v_A)^2 t + (v_B)^2 t - 2v_A v_B t \cos\theta}{\sqrt{(v_A t)^2 + (v_B t)^2 - 2v_A v_B t^2 \cos\theta + r^2}}$$

$$= \frac{v_B^2 t}{\sqrt{(v_B t)^2 + r^2}}.$$

(This is the same form as **Eq. (3)** in **Chapter 4** in the main body.)

If $v_B = 0$ in **Eq. (A1-12)**, (This is when B stays at point O_B.)

$$Y' = \frac{(v_A)^2 t + (v_B)^2 t - 2v_A v_B t \cos\theta}{\sqrt{(v_A t)^2 + (v_B t)^2 - 2v_A v_B t^2 \cos\theta + r^2}}$$

$$= \frac{v_A^2 t}{\sqrt{(v_A t)^2 + r^2}}.$$

(This is the same form as **Eq. (3)** in **Chapter 4**.)
Therefore, **Eq. (A1-12)** is correct.

(3) If $t = 0$ in **Eq. (A1-12)**,

$$Y' = \frac{(v_A)^2 t + (v_B)^2 t - 2v_A v_B t \cos\theta}{\sqrt{(v_A t)^2 + (v_B t)^2 - 2v_A v_B t^2 \cos\theta + r^2}} = \frac{0}{r} = 0.$$

That is, the relative speed of A with respect to B at the time $t = 0$ is zero (0). At his moment, the observation angle becomes 90° (see **Fig. A1-5**).

I refer to distance r ($=O_A O_B$) as "*r-factor*" in this case of 3-D situation. r is the shortest distance between two lines of motion of two objects in a 3-D situation (confer the definition of *r*-factor in 2-D situation in **Chapters 5** and **6**).

What if r is very large (∞)?
If $r = \infty$ in **Eq. (A1-12)**,

$$\lim_{r \to \infty} \frac{(v_A)^2 t + (v_B)^2 t - 2v_A v_B t \cos\theta}{\sqrt{(v_A t)^2 + (v_B t)^2 - 2v_A v_B t^2 \cos\theta + r^2}} = 0.$$

This means that if r is infinite (∞), the relative speed between the two objects is zero even when observation angle is not 90° as long as t (time) is not large (∞) enough to cancel the effect of $r = \infty$. This means that *the fourth law of relative speed* (see **Chapter 6**) holds also in 3-D situations. This is the principle that fixed stars look stationary (see the secret of fixed stars in **Chapter 6**). When viewed from a fixed star, our earth (and even the Galaxy, to which our earth belongs,) looks stationary. **Eq. A1-12** is a *special case* because two objects A and B start from A_0 and B_0 respectively at the same time. See the general case in **Exercise A1-6**. ♦

Exercise A1-6

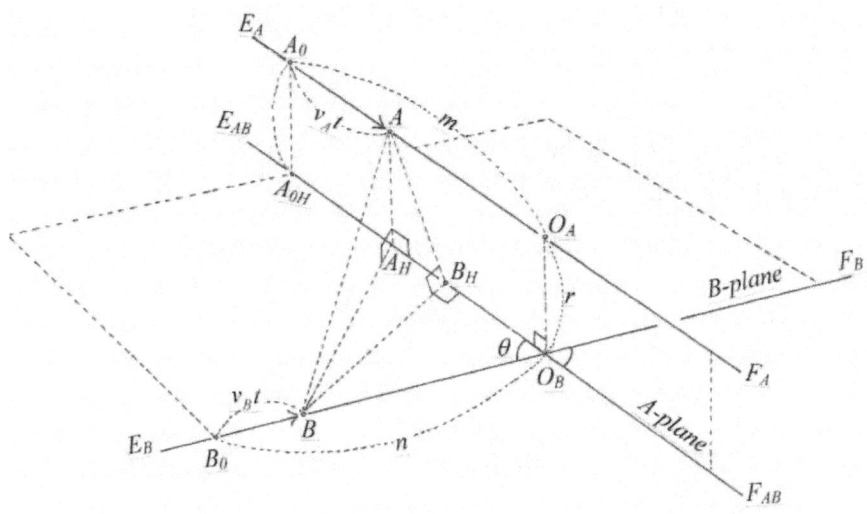

Fig. A1-6

In the above diagram, two straight lines $E_A F_A$ and $E_B F_B$ are not in the same plane. Point O_A is on straight line $E_A F_A$, and point O_B is on straight line $E_B F_B$. $O_A O_B$ is the shortest distance that connects the two straight lines $E_A F_A$ and $E_B F_B$. Straight line $E_{AB} F_{AB}$, which contains O_B, is parallel to $E_A F_A$. Now, $E_{AB} F_{AB}$ is the orthogonal projection of $E_A F_A$ on B-plane, which contains two straight lines $E_B F_B$ and $E_{AB} F_{AB}$. A-plane, which contains two parallel lines $E_A F_A$ and $E_B F_B$, is perpendicular to B-plane. Intersection angle $F_{AB} O_B E_B = \theta$. Point A_0 is distance m from O_A. Object A started from A_0 at time $t = 0$ and moves along straight line $E_A F_A$ in the direction $E_A \to F_A$ at the speed v_A. That is, $A_0 A = v_A t$. Point B_0 is distance n from O_B. Object B started from B_0 at time $t = 0$ and moves along straight line $E_B F_B$ in the direction $E_B \to F_B$ at the speed v_B. That is, $B_0 B = v_B t$. A_H is the point at which the perpendicular from A to $E_{AB} F_{AB}$ meets $E_{AB} F_{AB}$. Since A moves, A_H moves accordingly along $E_{AB} F_{AB}$ at the speed v_A. $O_A O_B = A A_H = r$.

Find the following:

(1) The equation that represents distance AB with respect to time t.

(2) The equation that represents the speed of A relative to B.

Solution:

Since $E_{AB}F_{AB}$ is the orthogonal projection of $E_A F_A$ on B-plane, angle $AA_H B$ = angle $AB_H B = 90°$.

(1) Let distance AB be Y. From right triangle ABB_H,

$$AB = Y = \sqrt{(AB_H)^2 + (BB_H)^2}$$

$$= \sqrt{(A_H B_H)^2 + (AA_H)^2 + (BO_B \sin\theta)^2}.$$

(Note: $A_H B_H = A_{0H} O_B - A_{0H} A_H - B_H O_B =$

$m - v_A t - (n - v_B t)\cos\theta$, $AA_H = r$, and

$BO_B \sin\theta = (B_0 O_B - B_0 B)\sin\theta = (n - v_B t)\sin\theta$. Therefore,

$$Y = \sqrt{[(m - v_A t) - (n - v_B t)\cos\theta]^2 + r^2 + [(n - v_B t)\sin\theta]^2}$$

$$= \sqrt{(m - v_A t)^2 + (n - v_B t)^2 - 2(m - v_A t)(n - v_A t)\cos\theta + r^2}. \quad \textbf{(A1-13)}$$

Eq. **(A1-13)** represents distance AB at time t.

(2) The derivative of **Eq. (A-13)** with respect to time t is

$$Y' =$$

$$\frac{v_A(v_A t - m) + v_B(v_B t - n) + [v_A(n - v_B t) + v_B(m - v_A t)]\cos\theta}{\sqrt{(m - v_A t)^2 + (n - v_B t)^2 - 2(m - v_A t)(n - v_B t)\cos\theta + r^2}}. \quad \textbf{(A1-14)}$$

Eq. **(A1-14)** is the speed equation of A relative to B.

How do we know whether **Eq. (A1-14)** is correct?

We can prove **Eq. (A1-14)** by replacing both m and n with 0 (zero) to see whether the result would be the same as **Eq. (A1-12)** in **Exercise A1-5**.

If $m = 0$ and $n = 0$ in **Eq. (A1-14)**,

Appendix-1

$$Y' = \frac{(v_A)^2 t + (v_B)^2 t - 2v_A v_B t \cos\theta}{\sqrt{(v_A t)^2 + (v_B t)^2 - 2v_A v_B t^2 \cos\theta + r^2}}.$$

[This is the same form as **Eq. (A1-12)**.]

Therefore, **Eq. (A1-14)** is correct.

Note: We have derived **Eq. (A1-14)** by using right triangle ABB_H. We can find the same result by using right triangle ABA_H.

What if r is very large (∞)?

If $r = \infty$ in **Eq. (A1-14)**,

$$\lim_{r \to \infty} \frac{v_A(v_A t - m) + v_B(v_B t - n) + [v_A(n - v_B t) + v_B(m - v_A t)]\cos\theta}{\sqrt{(m - v_A t)^2 + (n - v_B t)^2 - 2(m - v_A t)(n - v_B t)\cos\theta + r^2}}$$

$= 0$.

This means that if r (= the distance between two lines of motion of two objects in a 3-D situation) is infinite (∞), the relative speed between the two objects is zero even when observation angle is not 90° as long as t (time) is not large (∞) enough to cancel the effect of $r = \infty$. This means that ***the fourth law of relative speed*** (see **Chapter 6**) holds also in 3-D situations. This is the principle that fixed stars look stay put (see the secret of fixed stars in **Chapter 6**).

Ex A-5 and **Ex A-6** are the cases in which the involved objects are in linear motion. But in reality, the earth and fixed stars are in curvilinear motion in space. Even if the earth and a fixed star are in curvilinear motion, the instantaneous relative speed between the two is determined by the two tangent lines (straight lines) at a given motion. Therefore **Ex A1-5** and **A1-6** are valid in reality. **Ex A1-5** is a special case, and **A1-6** is a general case. ◆

APPENDIX-2
Four Laws of Relative Speed

The First Law of Relative Speed

In a 1-D situation, the relative speed between two involved objects is the difference between the proper speeds (or actual speeds) of the two objects ($= |v_1 - v_2|$) (see **Chapter 4**).

The Second Law of Relative Speed

In a 2-D or 3-D situation, the relative speed between the two involved objects is not simply the difference between the actual speeds of the two objects ($\neq v_1 \sim v_2$). In this case we can find the relative speed between the two objects by using calculus or trigonometry (cosine effect) or speed detector (see **Chapter 4**).

The Third Law of Relative Speed

At the moment the *observation angle* is 90°, the relative speed between two involved objects is zero (0) regardless of the types of motion or magnitudes of the two objects (see **Chapter 4**).

Fourth Law of Relative Speed

If r-factor (see **Chapter 5**) is very large (∞), the relative speed between two involved objects is zero (0) even when the observation angle is not 90° (see **Chapter 6**). ♦

APPENDIX-3

Einstein—His Religion, Philosophy, and Morality

Einstein's religion, philosophy, and political ideology have little to do with his theory of relativity. But peeping into his inner world of non-physical science is not only interesting but also helpful in understanding the depth of his intelligence and outlook on the world.

I got the related information (Einstein's words, anecdotes, etc.) mainly from Einstein's biography *Einstein/ His Life and Universe* (by Walter Isaacson, Simon & Schuster Paperbacks, 2007). Isaacson, the author, is a huge admirer of Einstein. However, unlike most other biographies, Isaacson's book includes a lot of episodes of Einstein that seem rather negative or detrimental to the fame of Einstein. In this respect, Isaacson's book is quite honest and fair in dealing with the icon of the 20th century.

Unlike the original intention of Isaacson, I have found many clues in Isaacson's book that Einstein not only made many mistakes in physical science but he also made many mistakes in his political ideology, philosophy, and religion. For example, Einstein's opinion about god and religion was inconsistent or ambiguous; he did not make clear the difference among theism, deism, and atheism. As a result, theists, deists, and atheists argue that Einstein is their side respectively. Einstein is a huge icon for both theists and atheists. This is unambiguously a comedy.

Einstein came to America in 1933 somewhat as a political refugee (from Nazi Germany) and became a US citizen in 1940. Throughout his life in America, until he died in 1955, Einstein's political ideas, articles, and public activities were dangerous to the freedom and interest of America and free world. Einstein was a staunch socialist; he courageously advocated the interest of communist Soviet. He made friends with Soviet spy Margarita Konencova, whose mission was to influence American scientists.

Einstein was not only the killer of physics but also he was trying to kill the freedom and prosperity of America, who had saved him from Nazi Germany and made him a prophet or god of science. Many patriotic American citizens in his time vehemently hated Einstein because his blatant and repeated pro-Soviet and anti-American activities exasperated them. But his fame as a world-renowned scientist managed to save him from his anti-American sin (treason).

The words in quotation marks in the following citations are Einstein's if not indicated otherwise. The page numbers included are of Isaacson's book *Einstein/ His Life and Universe*. The comment in the brackets after each citation is mine.

---

1. "Try and penetrate with our limited means the secrets of nature and you will find that, behind all the discernible laws and connections, there remains something subtle, intangible and inexplicable. Veneration for this force beyond anything that we can comprehend is my religion. To that extent I am, in fact, religious" (pp. 384-385).

[Here Einstein is a rather deist. Throughout his adult life, his opinion or position in relations with traditional religion was unclear and inconsistent.]

2. As a child, Einstein had gone through an ecstatic religious phase, and then rebelled against it. For the next three decades, he tended not to pronounce much on the topic. But around the time he turned 50, he began to articulate more clearly—in various essays, interviews, and letters—his deepening appreciation of his Jewish heritage and, somewhat separately, his belief in, albeit a rather impersonal, deistic concept of (p. 385).

[Einstein's was deistic which traditional religions object.]

3. Viereck began by asking Einstein whether he considered himself a

Appendix-3

German or a Jew. "It is possible to be both,' replied Einstein. "Nationalism is an infant disease, the measles of mankind" (p. 386).

[The reason he hated nationalism is that he thought that nationalism is the cause of war. Einstein believed in "world government" instead.]

4. "We Jews have been too eager to sacrifice our idiosyncrasies in order to conform". "As a child I received instruction both in the Bible and in the Talmud. I am a Jew, but I am enthralled by the luminous figure of the Nazarene" (p. 386).

[Einstein means that he respects Jesus and Christianity. Traditionally, Judaism and Christianity have been the enemies of each other. Many modern religionists say they respect the religions of other people. They say so not because they really respect the religions of other people but because they know that they cannot beat the religions of other people right now. This is a conditional or tactical truce.]

5. "I am not an atheist. The problem involved is too vast for our limited minds. We are in the position of a little child entering a huge library filled with books in many languages. The child knows someone must have written those books. It does not know how. It does not understand the languages in which they are written. The child dimly suspects a mysterious order in the arrangement of the books but does not know what it is. That, it seems to me, is the attitude of even the most intelligent human being toward. We see the universe marvelously arranged and obeying certain laws but only dimly understand these laws" (p. 386).

[Einstein is sometimes atheistic or sometimes theistic or sometimes deistic or sometimes all of these at the same time; We cannot tell what category he belonged to.]

6. "I am a determinist. I do not believe in free will. Jews believe in free will. They believe that man shapes his own life. I reject that doctrine. In that respect I am not a Jew" (p. 387).

[Judeo-Christian-Islamic religions are basically deterministic. These religions believe that everything is foretold in their scriptures. Prophets were those who revealed the secrets (predetermined plans) of.]

7. "I am fascinated by Spinoza's pantheism, but I admire even more his contribution to modern thought because he is the first philosopher to deal with the soul and body as one, and not two separate things" (p. 387).

[Spinoza was a deist. Judeo-Christian religions are basically dualism which means that the soul and body are different entities. Einstein's philosophy was monism.]

8. "I am enough of an artist to draw freely on my imagination. Imagination is more important than knowledge. Knowledge is limited. Imagination encircles the world" (p. 387).

[Imagination is important. But imagination that lacks rational knowledge is useless or harmful at times. Einstein's relativity is the product of imagination (theoretical study) that has no background of objective observation or experiment. Einstein's trick was to rationalize his theory with abstruse math so as to make the calculation is approximately same as that of classical physics. The person who works freely on his imagination is not a determinist. Determinism and freedom of thought do not mix. Was Einstein predetermined by to invent false physics?]

9. "I do not believe in immortality. And one life is enough for me" (p. 387).

[The belief in immortality is the core of Judeo-Christian religions. Einstein explicitly denies traditional religions.]

10. "I cannot conceive of a personal who would directly influence the actions of individuals or would sit in judgment on creatures of his own creation" (p. 387).

[Einstein was against personal God which is the core of Judeo-Christian religions. In this respect, Einstein was not a Jew; he was a deterministic deist. But in general, most deists are not determinists.]

11. "My religiosity consists of a humble admiration of the infinitely superior spirit that reveals itself in the little that can comprehend about the knowable world. That deeply emotional conviction of the presence of a superior reasoning power, which is revealed in the incomprehensible universe, forms my idea of" (p. 388).

[Einstein admits the presence of the infinitely superior spirit or a superior reasoning power which traditional religions refer to as. In this respect, Einstein admits that he believes in.]

Appendix-3

12. To a girl in the sixth grade who asked Einstein, saying, "Do scientists pray?" Einstein answered, "Scientific research is based on the idea that everything that takes place is determined by laws of nature, and this holds for the actions of people. For this reason, a scientist will hardly be inclined to believe that events could be influenced by a prayer, i.e. by a wish addressed to a supernatural Being" (p. 388).

[Einstein articulately denies personal. Yet he is a determinist. Einstein was right in his saying that determinists need not pray. Most scientists are neither religionists nor determinists. Traditional religionists are determinists, but they do pray hoping they can have God alter the predetermined destinies.]

13. Boston's Cardinal William Henry O'Connell said, "I very seriously doubt that Einstein himself really knows what he is driving at. The outcome of this doubt and befogged speculation about time and space is a cloak beneath which hides the ghastly apparition of atheism" (p. 388).

[Cardinal O'Connell considers Einstein a dangerous atheist. Relativity is basically against. I do think space and time were so created as to increase or shrink depending on the motion of human observers.]

14. "I believe in Spinoza's, who reveals himself in the lawful harmony of all that exists, but not in a who concerns himself with the fate and the doings of mankind." Einstein's response was not comforting to everyone. Some religious Jews, for example, noted that Spinoza had been excommunicated from the Jewish community of Amsterdam for holding these beliefs, and he had also been condemned by the Catholic Church for good measure (p. 388-389).

[Spinoza (1632-1677) maintained a pantheistic view which was untraditional for most Netherlands at that time. I like his view that "everything that exists, including individual men and women, is a part of." But his view upset his orthodox contemporaries--both Christians and Jews. The Jews denounced him and forced him to leave Amsterdam.]

Spinoza

(Source: *The World Book Encyclopedia*, 1979 ed.)

Spinoza, *spih NOH zuh*, **Baruch**, or Benedict (1632-1677), was a

> Dutch philosopher. He accepted Rene Descartes' idea that the universe is divided into mind and matter. But he saw, as Descartes did not, that if mind and matter are separate substances, they cannot interact. Spinoza decided that they are "attributes" of one substance,.., being infinite, has many attributes, but mind and matter are the only two that human minds can know.
>
> Among the consequences of this view is the following: everything that exists, including individual men and women, is a part of; in, quite literally, we live and move and have our being. This view upset Spinoza's orthodox contemporaries, both Christian and Jewish, because it was so different from their own. The Jews denounced him and forced him to leave Amsterdam. Spinoza actually was deeply religious, and in many respects was a mystic. He held that people's highest happiness consists in coming to understand and appreciate the truth that they are a tiny part of an all-inclusive, pantheistic.
>
> Spinoza was born in Amsterdam of Jewish parents. He broke with the Jewish faith after studying Descartes and Giordano Bruno. Spinoza prized independence and freedom of thought so much that he preferred to support himself by grinding lenses rather than accept a university professorship of financial aid.

15. "There are people who say there is no. But what makes me really angry is that they quote me for support of such views" (p. 389).

[Einstein's concept of was unclear for both believers and non-believers. This was the reason each of the two groups tried to use Einstein for its end. Einstein's ambiguity was the cause of this problem.]

16. "What separates me from most-so-called atheists is a feeling of utter humility toward the unattainable secrets of the harmony of the cosmos" (p. 389).

[Einstein proclaimed his relativity as the law of nature. Thus he led numerous people to misconception. Einstein was far from humility.]

17. "The fanatical atheists are like slaves who are still feeling the weight of their chains which they have thrown off after hard struggle. They are creatures who—in their grudge against traditional religion as the 'opium of the masses'—cannot hear the music of the spheres" (p. 390).

Appendix-3

[Atheism itself is not a religion. But Atheists believe in many other philosophies or ideologies. Many atheists believe in socialism. I regard socialism/communism as another religion or opium. I do not think Einstein ever heard the music of the spheres. At least his theory of relativity was not the music of the spheres.]

18. Einstein would later engage in an exchange on this topic with a U.S. Navy ensign he had never met. Was it true, the sailor asked, that Einstein had been converted by a Jesuit priest into believing in? That was absurd, Einstein replied. He went on to say that he `considered the belief in a who was a fatherlike figure to be the result of "childish analogies." Would Einstein permit him, the sailor asked, to quote his reply in his debates against his more religious shipmates? Einstein warned him not to oversimplify. "You may call me an agnostic, but I do not share the crusading spirit of the professional atheist whose fervor is mostly due to a painful act of liberation from the fetters of religious indoctrination received in youth," he explained. "I prefer the attitude of humility corresponding to the weakness of our intellectual understanding of nature and of our own being" (p.390).

[Einstein was not clear whether he was deist or atheist or theist.]

19. "The cosmic religious feeling is the strongest and noblest motive for scientific research" (p. 390).

[Throughout human history most religions have repressed or banned scientific research. Most religious leaders thought and think that science threatens or weakens the authority of their. Religionists believe that truth and solutions of all the problems in the world are already given in the scriptures.]

20. "Science can be created only by those who are thoroughly imbued with the aspiration toward truth and understanding. This source of feeling, however, springs from the sphere of religion" (p. 390).

[Einstein confuses the nature of science with that of religion.]

21. "Science without religion is lame, religion without science is blind" (p. 390).

[Science and religion are not complementary with each other; they are antagonistic with each other.]

22. "There is one religious concept that science cannot accept: a deity who could meddle at whim in the events of his creation or in the lives of his creatures" (pp. 390-391).

[Einstein was against personal. This means that he was against traditional religion.]

23. "The main source of present-day conflicts between the spheres of religion and of science lies in the concept of a personal" (p. 391).

[Einstein was right. A personal is the essence of religion. Science does not recognize personal.]

24. The belief in causal determinism, which was inherent in Einstein's scientific outlook, conflicted not only with the concept of a personal. It was also, at least in Einstein's mind, incompatible with human free will. Although he was a deeply moral man, his belief in strict determinism make it difficult for him to accept the idea of moral choice and individual responsibility that is at the heart of most ethical systems. Jewish as well as Christian theologians have generally believed that people have this free will and are responsible for their actions. They are even free to choose, as happens in the Bible, to defy God's commands, despite the fact that this seems to conflict with a belief that is all-knowing and all-powerful (p. 391).

[Most scientists do not believe in determinism. Though Einstein was not a traditional religionist, he believed in strict determinism, which is the thing of religion.]

25. Einstein, on the other hand, believed, as did Spinoza, that a person's actions were just as determined as that of billiard ball, planet, or star. "Human beings in their thinking, feeling and acting are not free but are as causally bound as the stars in their motions," Einstein declared in a statement to a Spinoza Society in 1932 (p. 391).

[Einstein was a mechanistic determinist.]

26. Do you believe, Einstein was once asked, that humans are free agents? "No, I am a determinist," he replied. "Everything is determined,

Appendix-3

the beginning as well as the end, by forces over which we have no control. It is determined for the insect as well as for the star. Human beings, vegetables, or cosmic dust, we all dance to the mysterious tune, intoned in the distance by an invisible player" (pp. 391-192).

[The "invisible player" is very personal of religionists. Einstein denied the probability or uncertainty principle of his peer scientists. He said that does not play dice. Einstein said that the universe maintains a constant size and shape. But later, he admitted that he had made a mistake, and he accepted the idea that the universe/space is expanding. In my opinion, space itself is neither expanding nor contracting.]

27. This attitude (Einstein's attitude of maintaining his strict determinism) appalled some friends, such as Max Born, who thought it completely undermined the foundations of human morality. "I cannot understand how you can combine an entirely mechanistic universe with the freedom of the ethical individual," he wrote Einstein. "To me a deterministic world is quite abhorrent. Maybe you are right, and the world is that way, as you say. But at the moment it does not really look like it in physics—and even less so in the rest of the world."

For Born, quantum uncertainty provided an escape from this dilemma. Like some philosophers of the time, he latched on to the indeterminacy that was inherent in quantum mechanics to resolve "the discrepancy between ethical freedom and strict natural laws." Einstein conceded that quantum mechanics called into question strict determinism, but he told Born he still believed in it, both in the realm of personal actions and physics (p. 392).

[Einstein was a strict and unnegotiable determinist.]

28. "I am compelled to act as if free will existed because if I wish to live in a civilized society I must act responsibly" (p. 392).

[At least Einstein acknowledged that citizens in the civilized society should act as free-will believers. Einstein was a hypocrite.]

29. "I know that philosophically a murderer is not responsible for his crime, but I prefer not to take tea with him" (p. 393).

[Einstein was perhaps pre-determined to prefer not to take tea with a murderer. But he could not be a chooser because he was a determinist.]

30. "The most important human endeavor is the striving for morality in our actions. Our inner balance and even our existence depend on it. Only morality in our actions can give beauty and dignity for life" (p. 393).

[It is surprising that a strict determinist emphasized the human striving for morality. Morality is based on free will and not on determinism.]

31. "The only salvation for civilization and the human race lies in the creation of world government. As long as sovereign states continue to have armaments and armament secrets, new world war will be inevitable" (pp. 487-488).

[World government does not work. Einstein's idea of world government was a ***super-super government*** which has the power to inspect and check the military means of all nations. Einstein dreamed of a super-big government that has the power to control or supervise all the nations of the world. Big government never fails to fail. The major cause of war is idealism, such as socialism, communism or religion, which promises a perfect world (paradise).

Einstein thought that it is not right that the secret of atomic bomb is monopolized by America alone. He was against the idea that America remains the only super power. Einstein hated American free capitalist society. Einstein favored or advocated Russia's socialism even in the days of his affluent living in America.]

32. Einstein fleshed out his proposals in a series of essays and interviews. The most important arose from an exchange of fan letters he had with Raymond Gram Swing, a commentator on ABC radio. Einstein invited Swing to visit him in Princeton, and the result was an article by Einstein, as told to Swing, in the November 1945 issue of the Atlantic called "Atomic War or Peace."

The three great powers—the United States, Britain, and Russia—should jointly establish the new world government, Einstein said in the article, and then invite other nations to join. Using a somewhat misleading phrase that was part of the popular debate of the time, he said that "the secret of the bomb" should be given to this new organization by Washington. The only truly effective way to control atomic arms, he

Appendix-3

believed, was by ceding the monopoly on military power to a world government (p.489).

[The secret of the nuclear bomb should be shared with Russia? At that time Russia did not get the secret of the bomb yet. Einstein's idea was more than naïve; his idea was dangerous. Einstein became a US citizen in 1940. He maintained pro-Soviet or pro-socialist sentiment throughout his life in America. Naturally, he condemned American freedom and capitalism.]

33. Einstein's efforts on behalf of arms control and his advocacy of world government in the late 1940s got him tagged as wooly-headed and naïve. Wooly-headed he may have been, at least in appearance, but was it right to dismiss him as naïve?

Most Truman administration officials, even those working on behalf of arms control, thought so. William Golden was an example. An Atomic Energy Commission staffer, who was preparing a report for Secretary of State George Marshall, went to Princeton to consult with Einstein. Washington needed to try harder to enlist Moscow in an arms control plan, Einstein argued. Golden felt he was speaking "with almost childlike hope for salvation and without appearing to have thought through the details of his solution." He reported back to Marshall, "It was surprising, though perhaps it should not have been, that, out of his métier of mathematics, he seemed naïve in the field of international politics. The man who popularized the concept of a fourth dimension could think in only two of them in considerations of World Government" (p. 493).

[Einstein was dangerously naïve and childlike in political matter.]

34. He (Einstein) disliked communist authoritarianism, but he did not see it as an imminent danger to American liberty. The greater danger, he felt, was rising hysteria about the supposed Red menace. When Norman Cousins, editor of the Saturday Review and the journalistic patron of America's internationalist intelligentsia, wrote a piece calling for international arms control, Einstein responded with a fan letter but added a caveat. "What I object to in your article is that you not only fail to oppose the widespread hysterical fear in our country of Russian aggression but actually encourage it," he said. "All of us should ask ourselves which of the two countries is objectively more justified in fearing the aggressive intentions of the other" (p. 495).

[Einstein believed that America was more dangerous and unjust a nation than Russia.]

35. As for the repression inside Russia, Einstein tended to offer only mild condemnations diluted by excuses. "It is undeniable that a policy of severe coercion exists in the political sphere," he said in one talk. "This may, in part, be due to the need to break the power of the former ruling class and to convert a politically inexperienced, culturally backward people into a nation well organized for productive work. I do not presume to pass judgment in these difficult matters" (p. 495).

[Einstein advocated the repression inside Russia as an inevitable measure to break the power of the former ruling class and establish a better system.]

36. "I do not approve of the interference by the Soviet government in intellectual and artistic matters. Such interference seems to me objectionable, harmful, and even ridiculous. Regarding the centralization of political power and the limitations of the freedom of action for the individual, I think that these restrictions should not exceed the limit demanded by security, stability, and the necessities resulting from a planned economy. An outsider is hardly able to judge the facts and possibilities. In any case it cannot be doubted that the achievements of the Soviet regime are considerable in the fields of education, public health, social welfare, and economics, and that the people as a whole have greatly gained by these achievements" (p. 496).

[Einstein did not understand where the suppression of the freedom of individuals of Russia came from. Instead, Einstein hailed the communistic achievements of socialist Russia. Einstein believed in the ideal world (paradise) socialism promises.]

37. He (Einstein) befriended many of the democratic socialist leaders in Britain and America, such as Bertrand Russell and Norman Thomas, and in 1945 he wrote an influential essay for the inaugural issue of the Monthly Review titled "Why Socialism?" In it he argued that unrestrained capitalism produced great disparities of wealth, cycles of boom and depression, and festering levels of unemployment. The system encouraged selfishness instead of cooperation, and acquiring wealth rather than serving others. People were educated for careers rather than for a love of work and creativity. And political parties became corrupted by political contributions from owners of great capital.

These problems could be solved, Einstein argued in his article, through a socialist economy, if it guarded against tyranny and centralization of power. "A planned economy, which adjusts production to the needs of the community, would distribute the work to be done among all those able to work and would guarantee a livelihood to every man, woman, and child," he wrote. "The education of the individual, in addition to promoting his own innate abilities, would attempt to develop in him a sense of responsibility for his fellow-men in place of the glorification of power and success in our present society" (p. 504).

[Einstein believed in socialism/communism; he believed in planned economy. He regarded capitalism evil. Einstein parroted the propaganda from Moscow. Socialists in free world overlooked the reality that governmental or bureaucratic corruptions are more serious in socialist states than in free capitalist states. Socialism, big government, thorough corruption, economic failure, death of freedom.... these are inseparable friends; no one or even can separate any of these from the rest.]

38. "I am a militant pacifist" (p. 376).

[True religionists are militant pacifists and militant war-makers at the same time. The energy and passion of religionists come from the aspiration of achieving their attractive goal—paradise and everlasting peace for all humanity.

Pacifism does not prevent war from happening. Empirically speaking, pacifists usher in wars rather than peace. Arthur Neville Chamberlain, the Prime Minister of the United Kingdom (1937 – 1940), is known for his appeasement foreign police in relation with Nazi Germany. Though he was a peace-lover, his naiveté or wishful thinking helped Hitler start World War II. Peace-lovers are frequently the useful idiots for war-makers.

It is not that religionists are evil people. The problem is their intention is too good. But too good an intention without sufficient reason brings too bad results. They realize a hell instead. The road to hell is paved with good intention.]

39. The socialist leader Norman Thomas tried to convince Einstein that pacifism could not occur without radical economic reforms. Einstein disagreed. "It is easier to win over people to pacifism than to socialism. We should work first for pacifism, and only later for socialism" (p. 375).

[Both Norman Thomas and Einstein were socialists; they differ in methodology of achieving socialist revolution. Einstein believes that pacifism is the first-and-wise step to socialism. Einstein was right. It is wiser strategy for socialists to advocate pacifism more loudly than socialism. Many American socialists believe that Einstein was a man of good conscience and his socialism was humane socialism.]

40. Einstein wrote a private letter asking that Julius and Ethel Rosenberg, who had been convicted of turning over atomic secrets to the Soviets, be spared the death penalty. He had avoided making any statements about the case, which had divided the nation with a frenzy seldom seen before the advent of the cable-TV age. Instead, he sent the letter to the judge, Irving Kaufman, with a promise not to publicize it. Einstein did not contend that the Rosenbergs were innocent. He merely argued that a death penalty was harsh in a case where the facts were murky and the outcome was driven more by popular hysteria than objectivity.

In a reflection of the tenor of the time, Judge Kaufman took the private letter and turned it over to the FBI. Not only was it put into Einstein's file, but it was investigated to see if it could be construed as disloyalty. After three months, a report was sent to Hoover saying no further incriminating evidence had been found, but the letter remained in the file.

When Judge Kaufman went ahead and imposed a death penalty, Einstein wrote to President Harry Truman, who was about to leave office, to ask him to commute the sentence. He drafted the letter first in German and then in English on the back of a piece of scrap paper that he had filled with a variety of equations that apparently, given how they trail off, led to nothing. Truman bucked the decision to incoming President Eisenhower, who allowed the executions to proceed.

Einstein's letter to Truman was released publicly, and the New York Times ran a front-page story headlined "Einstein Supports Rosenberg Appeal." More than a hundred angry letters swept in from across the nation. "You need some common sense plus some appreciation for what

Appendix-3

America has given you," wrote Marian Rawles of Portsmouth, Virginia. "You place the Jew first and the United States second," said Green, serving in Korea: "You evidently like to see our GI's killed. Go to Russia or back where you came from, because I don't like Americans like you living off this country and making un-American statements."

There was not as many positive letters, but Einstein did have a pleasant exchange with the liberal Supreme Court Justice William O. Douglas, who had unsuccessfully tried to stop the executions. "You have struggled so devotedly for the reaction of a healthy public opinion in our troubled time," Einstein wrote in a note of appreciation. Douglas sent back a handwritten reply: "You have paid me a tribute which brightens the burdens of this dark hour—a tribute I will always cherish" (pp. 525-526).

[Einstein believed that as long as sovereign states continue to have armaments and armament secrets, new world war would be inevitable. Einstein must have thought that the nuclear bomb secrets of America would lead to a new world war. Then sharing such secrets with Russia was good, according to Einstein's *World-Government* ideology.]

41. Many of the critical letters asked Einstein why he was willing to speak out for the Rosenbergs but not for the nine Jewish doctors whom Stalin had put on trial as part of an alleged Zionist conspiracy to murder Russian leaders. Among those who publically challenged what they saw as Einstein's double standard were the publisher of the *New York Post* and the editor of the *New Leader*.

Einstein agreed that the Russian actions should be denounced. "The perversion of justice which manifests itself in all the official trials staged by the Russian government deserves unconditional condemnations," he wrote. He added that individual appeals to Stalin would probably not do much, but perhaps a joint declaration from a group of scholars would help. So he got together with the chemistry Nobel laureate Harold Urey and others to issue one. "Einstein and Urey Hit Red's Anti-Semitism," the New York Times reported. (After Stalin died a few weeks later, the doctors were freed.)

On the other hand, he stressed in scores of letters and statements that Americans should not let the fear of communism cause them to surrender the civil liberties and freedom of thought that they cherished. There were a lot of domestic communists in England, but the people there did not get themselves whipped into frenzy by internal security investigations, he pointed out. Americans need not either (p. 526).

[The author (Walter Isaacson) gives readers the impression that Einstein's action was decisive in freeing the nine Jewish doctors in Russia. But it is not clear whether such positive outcome was due to Einstein's action or simply due to the death of Stalin. Stalin had been criticized by many Russian leaders, including his successor Nikita Khrushchev, even before Einstein took action. Khrushchev exonerated huge number of political prisoners.

Einstein seemed not to understand that the repression of human rights in Russian regime was related to communism. Einstein still advocated communism by saying that Americans should not be too allergic or sensitive to home-grown communists with the logic that British people were not so. The reason America had not become communist state was not that homegrown communists were harmless people but that many citizens resisted strongly against communists. It is the game of *number*. Had there been many socialists like Einstein in America and Britain, these two nations must have become socialist states.]

42. What is notable, in retrospect, about Einstein's FBI file is not all the odd tips it contained, but the one relevant piece of information that was completely missing. Einstein did in fact consort with a Soviet spy, unwittingly. But the FBI remained clueless about it.

The spy was Margarita Konenkova, who lived in Greenwich Village with her husband, the Russian realist sculptor Sergei Konenkova, mentioned earlier. A former lawyer who spoke five languages and had an engaging way with men, so to speak, her job as a Russian secret agent was to influence American scientists. She had been introduced to Einstein by Margot, and she became a frequent visitor to Princeton during the war.

Out of duty or desire, she embarked on an affair with the widowed Einstein. One weekend during the summer of 1941, she and some friends invited him to a cottage on Long Island, and to everyone's surprise he accepted. They packed a lunch of boiled chicken, took the train from Penn Station, and spent a pleasant weekend during which Einstein sailed

Appendix-3

on the Sound and scribbled equations on the porch. At one point they went to a secluded beach to watch the sunset and almost got arrested by a local policeman who had no idea who Einstein was. "Can't you read," the officer said, pointing to a no-trespassing sign. He and Konenkova remained lovers until she returned to Moscow in 1945 at age 51.

She succeeded in introducing him to the Soviet vice consul in New York, who was also a spy. But Einstein had no secret to share, nor is there any evidence that he had any inclination at all to help the Soviets in any way, and he rebuffed her attempts to get him to visit Moscow.

The affair and potential security issue came to light not because of any FBI sleuthing but because a collection of nine amorous letters written by Einstein to Konenkova in the 1940s became public in 1998. In addition, a former Soviet spy, Pavel Sudoplatov, published a rather explosive but not totally reliable memoir in which he revealed that she was an agent code-named "Lukas."

Einstein's letters to Konenkova were written the year after she left America. Neither she nor Sudoplatov, nor anyone else, ever claimed that Einstein passed along any secrets, wittingly or unwittingly (pp. 502-503).

[The missions of Russian spies are not only to steal the secrets of enemy nations but also indoctrinate celebrities with socialism. Yuri Bezmenov, a former KGB spy, who defected and came to America in 1970s, said that only 15 percent of all the money, effort, and manpower of KGB was executed to collect secrets of enemy nations; the rest—85 percent--was used to ideologically subvert (brainwash) as many important or influential people in the enemy nations as possible and let them form mild or friendly opinions about Russia and communism. I think Konenkova's mission to proselytize Einstein with pro-Soviet socialism was a piece of cake because Einstein had already been a pro-Soviet socialist. The chilling yet very informative confession interview of Yuri Bezmenov can be watched in the Youtube video at the following address:

http://www.youtube.com/watch?v=Cnf0I2dQ0i0, (running time 81m 29s)]

43. Even though he (Einstein) was honored for his nonconformity in the field of science, Einstein used the occasion to turn attention to the McCarthy-style investigations. For him, freedom in the realm of thought was linked to freedom in the realm of politics. "To be sure, we are concerned here with nonconformism in a remote field of endeavor," he said, meaning physics. "No Senatorial committee has as yet felt compelled to tackle the task of combating in this field the dangers that threaten the inner security of the uncritical or intimidated citizen."

Listening to his talk was a Brooklyn schoolteacher, William Frauenglass, who had a month earlier been called to testify in Washington before a Senate Internal Security Subcommittee looking into communist influence in high schools. He had refused to talk, and now he wanted Einstein to say whether he had been right.

Einstein crafted a reply and told Frauenglass he could make it public. "The reactionary politicians have managed to instill suspicious of all intellectual efforts," he wrote. "They are now proceeding to suppress the freedom of teaching." What should intellectuals do against this evil? "Frankly, I can only see the revolutionary way of non-cooperation in the sense of Gandhi's," Einstein declared. "Every intellectual who is called before one of the committee ought to refuse to testify."

Einstein's lifelong comfort in resisting prevailing winds made him serenely stubborn during the McCarthy era. At a time when citizens were asked to name names and testify at inquiries into their loyalty and that of their colleagues, he took a simple approach. He told people not to cooperate.

He felt, as he told Frauenglass, that this should be done based on the free guarantee of the First Amendment, rather than the "subterfuge" of invoking the Fifth Amendment's protection against possible self-incrimination. Standing up for the First Amendment was particularly a duty of intellectuals, he said, because they had a special role in society as preservers of free thought. He was still horrified that most intellectuals in Germany had not risen in resistance when the Nazi came to power.

When his letter to Frauenglass was published, there was an even greater public uproar than had been provoked by his Rosenberg appeal. Editorial writers across the nation pulled out all the stops for their denunciatory chords.

The *New York Times*: "To employ unnatural and il forces of civil disobedience, as Professor Einstein advises, is in this case to attack one evil with another. The situation which Professor Einstein rebels against certainly needs correction, but the answer does not lie in defying the

Appendix-3

law."

The *Washington Post*: He has put himself in the extremist category by his irresponsible suggestion. He has proved once more that genius in science is no guarantee of sagacity of political affairs."

The *Philadelphia Inquirer*: "It is particularly regrettable when a scholar of his attainments, full of honors, should permit himself to be used as an instrument of propaganda by the enemies of the country that has given such a secure refuge... Dr. Einstein has come down from the stars to dabble in ideological politics, with lamentable results."

The *Chicago Daily Tribune*: "It is always astonishing to find that a man of great intellectual power in some directions is a simpleton or even a jackass in others."

The *Pueblo* (Colorado) *Star-Journal*: "He, of all people, should know better. This country protected him from Hitler."

Ordinary citizen wrote as well. "Look in the mirror and see how disgraceful you look without a haircut like a wild man and wear a Russian wool cap like a Bolshevik," said Sam Epkin of Cleveland. The anticommunist columnist Victor Lasky sent a handwritten screed: "Your most recent blast against the institutions of this great nation finally convinces me that, despite your great scientific knowledge, you are an idiot, a menace to this country" (pp. 527-528).

[I generally agree with all the angry comments poured onto Einstein except for the comment that Einstein was a great scientist or genius. Einstein deserves many titles; he was a socialist, militant pacifist, One-World believer, and most of all, physics killer.]

44. The foundation of that morality, he (Einstein) believed, was rising above the "merely personal" to live in a way that benefited humanity. There were times when he could be callous to those closest to him, which shows that, like the rest of us humans, he had flaws. Yet more than most people, he dedicated himself honestly and sometimes courageously to actions that he felt transcended selfish desires in order to encourage human progress and the preservation of human freedoms. He was generally kind, good-natured, gentle, and unpretentious. When he and Elsa left for Japan in 1922, he offered her daughters some advice on how to lead a moral life. "Use for yourself little," he said, "but give to others much" (p. 393).

[Einstein was praised by many admirers as a man of high standard of morality. But many people criticized Einstein for his coldness toward his family members and for his infidelity. Alok Jha, science correspondent of *The Guardian* and admirer of Einstein, defended Einstein's morality in *the Guardian* as follows:
(ource: http://www.theguardian.com/science/2006/jul/11/internationalnews)

Letters reveal relative truth of Einstein's family life

Documents show 20th century giant was generous, affectionate - and adulterous / Monday 10 July 2006

He was the 20th century's greatest scientist, his name synonymous with genius. But while Albert Einstein's theories are known and lauded the world over, insights into his private life are patchy and largely negative. He has been variously portrayed as a bad father, cruel to his wives and an adulterer.

But that view could now change. Spanning more than 3,500 pages, a newly released set of Einstein's personal correspondence provides new clues into the character of the Nobel Prize-winning scientist. He was open about his love affairs to his wife, lost much of his prize money in bad investments and was a much more devoted father than previously thought.

According to Hanoch Gutfreund of the Hebrew University in Jerusalem, who is chairman of the Albert Einstein Worldwide Exhibition, the new letters shatter myths that the great scientist was always cold towards his family.

"Anybody who wants to write a new biography of Einstein will have an additional resource to take into account. As a result of that, certain chapters in his life will now emerge in a slightly different light than before," said Prof Gutfreund.

Einstein became known as one of the greatest physicists of all time after publishing the theory of special relativity in 1905 and a theory of gravity known as general relativity in 1916. He also made significant contributions to quantum mechanics and cosmology. In 1921, he was awarded the Nobel Prize for physics and has since become most famous for his equation showing the relationship between mass and energy: $E = mc^2$.

Appendix-3

Einstein was married twice, to Mileva Maric from 1903 until 1919 and to his cousin Elsa from 1919 until her death in 1936. Previously released letters suggested that his first marriage was miserable, and that he cheated on Elsa with his secretary, Betty Neumann. Prof Gutfreund said that though Einstein's marriage to Elsa was best described as one of convenience, he wrote to her constantly, describing, among other things, his experiences touring and lecturing.

"The general concept from everything we knew before was that he was a poor father, that he did not meet his responsibility to his children and that he was quite cruel to his wife," he said.

When he wanted a divorce from his first wife, Einstein gave her the ultimatum that, if she wanted to remain with him and not grant him a divorce, then he expected her to serve him three meals a day in his room but not expect any intimacy in return. "From the documents we have now, a different picture emerges," said Prof Gutfreund. "He does show empathy and compassion."

There is evidence that he diverted part of his winnings from the 1921 Nobel Prize into providing for Mileva and his children. He invested the rest in Europe and America - and lost much of it during the Great Depression.

Einstein was surprisingly candid to Elsa about his extramarital affairs. Between the mid-1920s and his emigration to the US in 1933, there were several women in his life: a Margarete, an Estella, two women called Toni and an Ethel. He shared holidays with them, read books and attended concerts.

In a letter to Elsa, he said women were chasing him, showering him with unwanted attention.

But he was aware of his weaknesses. "He was not capable of long and stable relations with a woman and he actually expressed that in a letter to the son of a friend who died," said Prof Gutfreund. Einstein wrote: "What I admire in your father is that, for his whole life, he stayed with only one woman. This is a project in which I grossly failed, twice."

"If one talks about Einstein in love, his most consistent love from beginning to end was science," said Prof Gutfreund.

Another apparent difficulty for Einstein was his relationship to his schizophrenic son, Eduard. "He refers [in previously known letters] to Eduard as maybe it would have been better if he would not have been born," said Prof Gutfreund. However, in the new letters Einstein writes of his pleasure in receiving poems, pictures and notes from him. Einstein wrote to friends: "The more refined of my sons, the one I considered really of my own nature, was seized by an incurable mental illness."

Einstein was much closer to Elsa's daughter, Margot. He wrote: "I love her [Margot] as much as if she were my own daughter, perhaps even more so, since who knows what kind of brat she would have become [had I fathered her]."

The 1,300 letters, which span from 1912 to Einstein's death in 1955, have been in storage at the Hebrew University in Jerusalem, shielded from the public in accordance with Margot's request that they be locked away for 20 years after her death. Margot died in July 1986.

Though the letters do not concern Einstein's theories, he does mention his weariness at being continually associated with his work. "Soon I'll be fed up with the relativity," he wrote to Elsa. "Even such a thing fades away when one is too involved with it."

Extracts: 'Soon I'll be fed up with the relativity'

Albert Einstein wrote to his wife Elsa almost every day and often to his stepdaughter Margot

To Elsa, from Prague, January 8 1921

My lectures here ... are already behind me. This morning quartet - very beautiful, like old times. The first violin is played by a youth of 80 years! Soon I'll be fed up with the relativity. Even such a thing fades away when one is too involved with it ....

To Margot, from Oxford. May 8 1931

(Members of Einstein's extended family were used to his involvement with two or three women, but had complained about the new additions to his harem.)

This time I'm writing you because you are the most reasonable [member of the family], and the poor mother [Elsa is] already completely meschugge. It is true that M. followed me and her chasing after me is getting out of control. But firstly I could hardly avoid it, and secondly, when I see her, I will tell her that she should vanish immediately.... Out of all the dames I am in fact attached only to Mrs. L who is absolutely harmless.

To Elsa

Appendix-3

Mrs. M definitely acted according to the best Christian-Jewish ethics: 1) one should do what one enjoys and what won't harm anyone else; and 2) one should refrain from doing things one does not take delight in and which annoy another person. Because of 1) she came with me, and because of 2) she didn't tell you a word.

To Elsa from Kiel. June 11 1933

(Elsa managed the financial affairs. From the moment Einstein became famous, she recognised his handwritten manuscripts would be a source of income. This letter was written when Einstein was working on the improvement of the gyroscope compass for the Anschuetz Company. Hermann Anschuetz had provided him with an apartment where he was shielded from the public.)

I don't want to have the Warburgs bothered with my manuscript, and much less Haldane. I don't mind having it sold, but without molesting any prominent people. Thank goodness one cannot sell my skin during my lifetime ... Here there is blessed calm. No one is allowed to ... claim any rights on me. Anschuetz admires me for my abstaining from smoking, and I admire myself, too. In front of my window [are] trees and water, chirping birds. Nothing unexpected occurs, everything quiet and comfortable as if arranged for contemplative musing.

It seems that Alok Jha is not that successful in defending Einstein's morality.] ♦

Epilogue

The job of disproving relativity or understanding the fallacy of relativity is relatively easy as to be done by any ordinary high school students. One needs not earn a doctor's degree or master's degree in physics or math to disprove relativity. Earning a degree in modern physics is rather a poison that kills the rational thinking about physical reality. Relativity is *armchair sophistry* that defies the physical reality.

Einstein was the emperor without clothes. The belief in relativity is an idolatry that worships the naked emperor as the god of science. The belief in relativity is more than idolatry; it is a colossal organized crime against humanity, Nature and the Creator of the universe, if any.

I disclosed the names of some eminent incumbent physicists (relativists) in this book because if I do not do so no one would pay attention to my arguments. Physicists (both relativists and dissidents) have ignored my arguments for the past 13 years (2001-2014). Physicists (both relativists and dissidents) should start all over again from exploring what relative speed is.

The theory of relativity is the shame of humanity. I hope American lawyers and judges can handle the criminal nature of relativity in court and put an end to the shame of humanity. This is to save the face of America who has been one of the main proselytizers of relativity. But if American scholars and lawyers are unable to do so, I hope that this job will be done by other countries such as China or South Korea or Japan.

Author / August, 2014 ♦

References

Summary

[1] Albert Einstein, *Relativity (The Special and the General Theory)* (15th ed.), Three Rivers Press, New York, 1961, p. 15.

[2] Albert Einstein, *Relativity (The Special and the General Theory)* (15th ed.), Three Rivers Press, New York, 1961, p. 69.

[3] Walter Isaacson, *Einstein His Life and Universe* (Simon & Schuster Paperbacks, New York, 2008), p. 318.

[4] http://en.wikipedia.org/wiki/Emission_theory, "Emission Theory."

[5] Albert Einstein, *Relativity (The Special and the General Theory)* (15th ed.), Three Rivers Press, New York, 1961, pp. 158-159.

[6] Albert Einstein, *Relativity (The Special and the General Theory)* (15th ed.), Three Rivers Press, New York, 1961, pp. 104-107.

[7] Albert Einstein, *Relativity (The Special and the General Theory)* (15th ed.), Three Rivers Press, New York, 1961, pp. 29-30.

[8] Albert Einstein, *Relativity (The Special and the General Theory)* (15th ed.), Three Rivers Press, New York, 1961, pp. 75-79.

[9] Albert Einstein, *Relativity (The Special and the General Theory)* (15th ed.), Three Rivers Press, New York, 1961, p. 85.

[10] Albert Einstein, *Relativity (The Special and the General Theory)* (15th ed.), Three Rivers Press, New York, 1961, p. 77.

[11] Walter Isaacson. *Einstein, his Life and Universe*, Simon & Schuster Paperbacks, New York, 2007, pp. 107-139.

[12] Albert Einstein, *Relativity (The Special and the General Theory)* (15th ed.), Three Rivers Press, New York, 1961, pp. 75-79.

[13] Albert Einstein, *Relativity (The Special and the General Theory)* (15th ed.), Three Rivers Press, New York, 1961, pp. 88-91.

[14] Albert Einstein, *Relativity (The Special and the General Theory)* (15th ed.), Three Rivers Press, New York, 1961, p. *vii*.

[15] Albert Einstein, Relativity (The Special and the General Theory) (15th ed.), Three Rivers Press, New York, 1961, pp. 155-156.

[16] Albert Einstein, *Relativity (The Special and the General Theory)*

(15th ed.), Three Rivers Press, New York, 1961, p. 6.
[17] Arthur Beiser, *Concept of Modern Physics* (5th international ed.). McGraw-Hill, Inc., New York, 2003, p. 40.
[18] Arthur Beiser, *Concept of Modern Physics* (5th international ed.). McGraw-Hill, Inc., New York, 2003, p. 11.
[19] Md. Farid Ahmed et al. "A Review of One-Way and Two-Way Experiments to Test the Isotropy of the Speed of Light," Ontario, 2011. (http://arxiv.org/ftp/arxiv/papers/1011/1011.1318.pdf)
[20] Albert Einstein, *Relativity* (*The Special and the General Theory*) (15th ed.), Three Rivers Press, New York, 1961, p. 77.
[21] *The World Book Encyclopedia* (1979 ed.), World Book-ChildCraft International, Inc. 1979, Chicago, "Force."
[22] *The World Book Encyclopedia* (1979 ed.), World Book-ChildCraft International, Inc. 1979, Chicago, "Weight"
[23] Albert Einstein, *Relativity* (*The Special and the General Theory*) (15th ed.), Three Rivers Press, New York, 1961), p. 75-79.
[24] Albert Einstein, *Relativity* (*The Special and the General Theory*) (15th ed.), Three Rivers Press, New York, 1961, p. 89.
[25] Albert Einstein, *Relativity* (*The Special and the General Theory*) (15th ed.), Three Rivers Press, New York, 1961, pp. 75-79; 83-87; 88-91
[26] Walter Isaacson, *Einstein His Life and Universe*, Simon & Schuster Paperbacks, New York, 2008, pp. 255-258.
[27] http://en.wikipedia.org/wiki/Arthur_Eddington, "Arthur Eddington"
[28] Arthur Beiser, *Concept of Modern Physics* (5th international ed.), McGraw-Hill, Inc., New York, 2003, p. 33.
[29] http://www.einstein-online.info/)→http://www.einstein-online.info/ spotlights/ atombomb.
[30] http://www.bibliotecapleyades.net/esp_einstein.htm, "Einstein, Plagiarist of the Century."
[31] Christopher Jon Bjerknes, "A Theory of Einstein the Irrational Plagiarists." The Canberra Times September 19, 2006. (http://www.jewwatch.com/jew-leaders-einstein-hoax1.html .)
[32] Walter Isaacson, *Einstein His Life and Universe*, Simon & Schuster Paperbacks, New York, 2008, p. 262.

Chapter 1
[1] http://en.wikipedia.org/wiki/Speed_of_sound, "Speed of Sound."
[2] Albert Einstein, *Relativity* (*The Special and the General Theory*)

References

(15th ed.), Three Rivers Press, New York, 1961, p. 15.

[3] http://en.wikipedia.org/wiki/Proper_velocity, "Proper Velocity."

[4] Albert Einstein, *Relativity (The Special and the General Theory)* (15th ed.), Three Rivers Press, New York, 1961, pp. 85.

[5] Albert Einstein, *Relativity (The Special and the General Theory)* (15th ed.), Three Rivers Press, New York, 1961, p. 77.

[6] Walter Isaacson, *Einstein His Life and Universe*, Simon & Schuster (Paperbacks, New York, 2008), p. 318.

[7] Albert Einstein, *Relativity (The Special and the General Theory)* (15th ed.), Three Rivers Press, New York, 1961, pp. 88-91.

Chapter 2

[1] Albert Einstein, *Relativity (The Special and the General Theory)* (15th ed.), Three Rivers Press, New York, 1961, p. 163.

Chapter 4

[1] Byoung Ha Ahn, "The Speed of an Object with Respect to an Observer Who is Off the Line of Motion of the Object (2005)," pp. 89.

[2] Byoung Ha Ahn, "The Speed of an Object with Respect to an Observer Who is Off the Line of Motion of the Object (2006)," pp. 45.

Chapter 6

[1] *The World Book Encyclopedia* (1979 ed.), World Book-ChildCraft International, Inc. 1979, Chicago, "Fixed Star."

[2] Albert Einstein, *Relativity (The Special and the General Theory)* (15th ed.), Three Rivers Press, New York, 1961, p. 15.

Chapter 7

[1] Albert Einstein, *Relativity (The Special and the General Theory)* (15th ed.), Three Rivers Press, New York, 1961, p. 15.

[2] http://en.wikipedia.org/wiki/Proper_velocity, "Proper Velocity."

Chapter 8

[1] Byoung Ha Ahn, "The Speed of an Object with Respect to an Observer Who is Off the Line of Motion of the Object (2004)," pp. 38.

[2] http://www.copradar.com/preview/chapt2/ch2d1.html, "Cosine Effect Error."

[3] Byoung Ha. Ahn, *Relativity, A Mistake of the 20th Century* (2000, registered but not unpublished), pp. 1-8.

[4] Byoung Ha Ahn, "The Speed of an Object with Respect to an Observer Who is Off the Line of Motion of the Object (2006)," pp. 45

Chapter 13

[1] Albert Einstein, *Relativity (The Special and the General Theory)*

(15th ed.), Three Rivers Press, New York, 1961, pp. 88-91.

Chapter 14

[1] http://en.wikipedia.org/wiki/Global_Positioning_System, "Global Positioning System."

[2] Ronald R. Hatch, "Relativity and GPS -1," 1995" (http://www.worldsci.org/pdf/abstracts/abstracts_1783.pdf.).

[3] Barry Stringer, "Does the GPS System Rely upon Einstein's Relativity?," 2012.
(http://www.worldnpa.org/site/member/?memberid=2323&subpage=abstracts.)

Chapter 15

[1] Albert Einstein, *Relativity* (*The Special and the General Theory*) (15th ed.), Three Rivers Press, New York, 1961, pp. 29-31.

[2] Doug Marett, "The Sagnac Effect: Does it Contradict Relativity?," 2012.
(http://www.conspiracyoflight.com/SagnacRel/SagnacandRel.html.)

[3] Albert Einstein, *Relativity* (*The Special and the General Theory*) (15th ed.), Three Rivers Press, New York, 1961, pp. 27-28.

Chapter 16

[1] Albert Einstein, *Relativity* (*The Special and the General Theory*) (15th ed.), Three Rivers Press, New York, 1961, pp. 13-14.

[2] Albert Einstein, *Relativity* (*The Special and the General Theory*) (15th ed.), Three Rivers Press, New York, 1961, p. 77.

[3] Albert Einstein, *Relativity* (*The Special and the General Theory*) (15th ed.), Three Rivers Press, New York, 1961, p. 89.

Chapter 18

[1] Walter Isaacson, *Einstein His Life and Universe*, Simon & Schuster Paperbacks, New York, 2008, p. 318.

[2] Walter Isaacson, *Einstein His Life and Universe*, Simon & Schuster Paperbacks, New York, 2008, pp. 199-201.

[3] Albert Einstein, "On the Foundation of the General Theory of Relativity" (Annalen der Physik Mar. 6, 1918, CPAE 7: 4).

[4] Walter. Isaacson, *Einstein His Life and Universe*, Simon & Schuster Paperbacks, New York, 2008, p.198.

[5] Albert Einstein, *Relativity* (*The Special and the General Theory*) (15th ed.), Three Rivers Press, New York, 1961, p. vii.

[6] Albert Einstein, *Relativity* (*The Special and the General Theory*) (15th ed.), Three Rivers Press, New York, 1961, pp. 158-163.

References

[7] http://en.wikipedia.org/wiki/Emission_theory, "Emission Theory."

Chapter 19

[1] *The World Book Encyclopedia* (1979 ed.), World Book-ChildCraft International, Inc. 1979, Chicago, "Fixed Star."

[2] Albert Einstein, *Relativity (The Special and the General Theory)* (15th ed.), Three Rivers Press, New York, 1961, p. 13.

Chapter 20

[1] http://en.wikipedia.org/wiki/Inertia, "Inertia."

[2] *The World Book Encyclopedia*, 1979 ed., World Book-ChildCraft International, Inc. Chicago, 1979, "Gyroscope."

Chapter 21

[1] Md. Farid Ahmed et al. "A Review of One-Way and Two-Way Experiments to Test the Isotropy of the Speed of Light," Ontario, 2011. (http://arxiv.org/ftp/arxiv/papers/1011/1011.1318.pdf)

Chapter 22

[1] Md. Farid Ahmed et al. "A Review of One-Way and Two-Way Experiments to Test the Isotropy of the Speed of Light," Ontario, 2011. (http://arxiv.org/ftp/arxiv/papers/1011/1011.1318.pdf)

Chapter 23

[1] Arthur Beiser, *Concept of Modern Physics* (5th international ed.), McGraw-Hill, Inc., New York, 2003, p. 50.

[2] Walter Isaacson. *Einstein, his Life and Universe*, Simon & Schuster Paperbacks, New York, 2007, p. 107-139.

Chapter 24

[1] Arthurt Beiser, *Concept of Modern Physics* (5th international ed.), McGraw-Hill, Inc., New York, 2003, pp. 36-39.

[2] Arthur Beiser, *Concept of Modern Physics* (5th international ed.), McGraw-Hill, Inc., New York, 2003, p. 38.

Chapter 26

[1] https://en.wikipedia.org/wiki/Speed_of_light, "Speed of Light."

Chapter 27

[1] Arthur Beiser, *Concept of Modern Physics* (5th international ed.), McGraw-Hill, Inc., New York, 2003, p. 4.

[2] http://en.wikipedia.org/wiki/Michelson%E2%80%93Morley_experiment, "Morley Experiment."

[3] *The World Book Encyclopedia* (1979 ed.), World Book-ChildCraft International, Inc. 1979, Chicago, "Earth"

Chapter 28

[1] Albert Einstein, *Relativity (The Special and the General Theory)* (15th ed.), Three Rivers Press, New York, 1961, pp. 90-91.

[2] http://home.comcast.net/~xtxinc/MainPage.htm, "Albert Einstein the Incorrigible Plagiarist."

[3] Richard Moody, Jr "Einstein, Plagiarist of the Century." (Extracted from *Nexus Magazine* Volume 11, Number 1 (December 2003-January 2004) from Nexus Magazine Website recovered through WayBackMachine Website Spanish version). (http://www.bibliotecapleyades.net/esp_einstein.htm)

[4] Walter Isaacson, *Einstein His Life and Universe*, Simon & Schuster Paperbacks, New York, 2008, p. 318.

[5] Albert Einstein, *Relativity (The Special and the General Theory)* (15th ed.), Three Rivers Press, New York, 1961, pp. 155-178.

Chapter 29

[1] Arthur Beiser, *Concept of Modern Physics* (5th international ed.). McGraw-Hill, Inc., New York, 2003, pp. 17-19.

[2] Albert Einstein, *Relativity (The Special and the General Theory)* (15th ed.), Three Rivers Press, New York, 1961, p. 77.

[3] Arthur Beiser, *Concept of Modern Physics* (5th international ed.), McGraw-Hill, Inc., New York, 2003, p. 19.

Chapter 30

[1] Arthur Beiser, *Concept of Modern Physics* (5th international ed.), McGraw-Hill, Inc., New York, 2003, pp. 15-17.

[2] Arthur Beiser, *Concept of Modern Physics* (5th international ed.), McGraw-Hill, Inc., New York, 2003, pp. 17-19.

Chapter 31

[1] Arthur Beiser, *Concept of Modern Physics* (5th international ed.), McGraw-Hill, Inc., New York, 2003, pp. 5-9.

Chapter 32

[1] http://en.wikipedia.org/wiki/Mass%E2%80%93energy_equivalence).

[2] http://www.einstein-online.info/)→http://www.einstein-online.info/spotlights/atombomb.

Chapter 33

[1] Albert Einstein, *Relativity (The Special and the General Theory)* (15th ed.), Three Rivers Press, New York, 1961, pp. 49-51.

Chapter 34

[1] Albert Einstein, *Relativity (The Special and the General Theory)*

References

(15th ed.), Three Rivers Press, New York, p. 77.

Chapter 35

[1] Albert Einstein, *Relativity (The Special and the General Theory)* (15th ed.), Three Rivers Press, New York, pp. 75-79.

[2] *The World Book Encyclopedia* (1979 ed.), World Book-ChildCraft International, Inc. 1979, Chicago, "Mass"

Chapter 36

[1] Albert Einstein, *Relativity (The Special and the General Theory)* (15th ed.), Three Rivers Press, New York, pp. 75-79.

[2] Albert Einstein, *Relativity (The Special and the General Theory)* (15th ed.), Three Rivers Press, New York, pp. 32-33.

[1] Albert Einstein, *Relativity (The Special and the General Theory)* (15th ed.), Three Rivers Press, New York, p. 75.

Chapter 37

[1] Arthur Beiser, *Concept of Modern Physics* (5th international ed.). McGraw-Hill, Inc., New York, 2003, pp. 32-35.

[2] Albert Einstein, *Relativity (The Special and the General Theory)* (15th ed.), Three Rivers Press, New York, p. 85.

[3] Albert Einstein, *Relativity (The Special and the General Theory)* (15th ed.), Three Rivers Press, New York, p. 77.

[4] Albert Einstein, *Relativity (The Special and the General Theory)* (15th ed.), Three Rivers Press, New York, pp. 88-91.

Chapter 38

[1] Arthur Beiser, *Concept of Modern Physics* (5th international ed.). McGraw-Hill, Inc., New York, 2003, pp. 83-86.

Chapter 39

[1] Albert Einstein, *Relativity (The Special and the General Theory)* (15th ed.), Three Rivers Press, New York, pp. 75-79.

[2] *The World Book Encyclopedia* (1979 ed.), World Book-ChildCraft International, Inc. 1979, Chicago, "Force"

Chapter 40

[1] Albert Einstein, *Relativity (The Special and the General Theory)* (15th ed.), Three Rivers Press, New York, pp. 88-91.

Chapter 41

[1] Albert Einstein, *Relativity (The Special and the General Theory)* (15th ed.), Three Rivers Press, New York, pp. 101-103.

[2] Albert Einstein, *Relativity (The Special and the General Theory)* (15th ed.), Three Rivers Press, New York, pp. 104-107.

[3] Arthur Beiser, *Concept of Modern Physics* (5th international ed.), McGraw-Hill, Inc., New York, 2003, p. 33.

Chapter 42

[1] Walter Isaacson, *Einstein His Life and Universe*, Simon & Schuster Paperbacks, New York, 2008, pp. 256-258.

[2] *The World Book Encyclopedia* (1979 ed.), World Book-ChildCraft International, Inc. 1979, Chicago, "Sun"

[3] *The World Book Encyclopedia* (1979 ed.), World Book-ChildCraft International, Inc. 1979, Chicago, "Refraction."

[4] Albert Einstein, *Relativity (The Special and the General Theory)* (15th ed.), Three Rivers Press, New York, pp. 146-147.

[5] http://en.wikipedia.org/wiki/Arthur_Eddington, "Arthur Eddington"

[6] ttp://www.kps.or.kr/home/kor/morgue/physicist/physicist_7.asp?globalmenu=6&localmenu=1&physicistpagenum=7

[7] Walter Isaacson, *Einstein His Life and Universe*, Simon & Schuster Paperbacks, New York, 2008, pp. 260-262.

[8] http://scienceworld.wolfram.com/biography/Eddington.html.

[9] James Coleman, "Relativity for the Layman" (translation) (Damoon, Seoul, 1994), p.156.

Chapter 43

[1] Arthur Beiser, *Concept of Modern Physics* (5th international ed.). McGraw-Hill, Inc., New York, 2003, p. 50.

[2] Walter Isaacson, *Einstein His Life and Universe*, Simon & Schuster Paperbacks, New York, 2008, p. 342.

[3] Walter Isaacson, *Einstein His Life and Universe*, Simon & Schuster Paperbacks, New York, 2008, pp. 336-344.

[4] Walter Isaacson, *Einstein His Life and Universe*, Simon & Schuster Paperbacks, New York, 2008, p. 466.

Index

A
$a = F/m$ 251
Absolute coordinate system 38, 84
Absolute magnitude 54
Absolute motion/speed 36, 61, 131, 149, 161
Absolute space 24, 61, 148, 160
Absolute theory 28
Abstract math 5
Accelerated motion 61, 109
Actual (real) force 252
Actual accelerated motion 31
Actual change 22
Actual motion/speed 17, 23, 57, 84, 92, 110, 149
Actual position 260
Ahn's light clock experiment 216
Almighty machine 21
Amorphous entity 5
Angular speed 61, 112, 115
Anisotropy of light speed 128
Apparent phenomenon 251
Apparent position 260
Apparition of atheism 309
Armchair discussion 21
Armchair scholar 21
Armchair sophistry 328
Arthur Beiser 37, 45, 173, 176, 187, 199,

Artificial acceleration 251
Artificial gravity 43
Artificial satellite 112
Atheist 307
Atomic bomb 221, 222
Average relative speed 21, 58
Average speed 39, 58, 81

B
Barry Springer 120
Bertrand Russell 316
Bible 307
Big Bang 17, 147, 183
Binding energy 223
Biological clock 199, 202
Black hole 17
Braggart 275
Bruno 310

C
Capitalism 315
Centrifugal force 40, 64, 253
Cheater 275
Chest thought experiment 29, 41, 247
Chromosphere 265
Circular reasoning 16, 33, 170, 177
Clock paradox 210
Collision course 92, 93
Comedy 305
Comparison photo 269
Concentric circles 37, 161
Continuous acceleration 26

Continuous jerk 236
Coordinate system 21, 63
4-D coordinate system 22
Coriolis effect (force) 133, 134
Corona 266
Corpuscular model 27
Corpuscular theory 146
Corpuscular theory 27, 45
Cosine effect 90
Cult phenomenon 48
Curvature of space-time 258
Curvilinear speed 60
Curvilinear motion 12, 21, 243, 303

D
Dark energy 17
Dark matter 17
Deflection of light 238, 260
Deist 306
Deluder 275
Delusion 5, 15, 136, 275
Derivative 72
Descartes 35, 142, 309
Determinism 312
Determinist 307
Dimensional situation 19, 55, 66
  1-D situation 11, 12, 19, 55, 66
  2-D situation 12, 19, 55, 66
  3-D situation 12, 19, 55, 66
Direction 54
Disc thought experiment 40, 253
Disciple 275
Distance 53
Distance contraction 13
Distance elongation 13
Donald Sawicki 92
Doppler Effect 38, 94, 342
Dualism 307

E
Eccentric circles 161
Eccentric circles 37
Eddingon's eclipse observation 43
Eddington 43, 242, 259
Eddington number 275
$E = mc^2$ 46, 219, 325
Einstein number 275
Emission theory 146
Emperor without clothes 328
Engels 275
Environment 21
Environmental reference body 21, 63
Eq. (1) 71
Eq. (2) 72
Eq. (3) 73
Eq. (4) 91
Eq. (5) 102
Eq. (6) 102
Eq. (7) 107
Eq. (8) 107
Eq. (8) 110
Eq. (9) 110
Eq. (A1-1) 286
Eq. (A1-10) 296
Eq. (A1-11) 299
Eq. (A1-12) 299
Eq. (A1-13) 302
Eq. (A1-14) 302
Eq. (A1-2) 286
Eq. (A1-3) 290
Eq. (A1-4) 290
Eq. (A1-5) 290
Eq. (A1-6) 290
Eq. (A1-7) 293
Eq. (A1-8) 293
Eq. (A1-9) 296
Equality of gravitational mass and inertial mass, The 39
Equivalence principle 40

Index

Ether 24, 26, 58, 128, 147, 183
Extension of Ether theory 24, 58
Euclidian geometry 252
Euclidian space-time 28
Experimental error 65
Eureka moment 76
Extension of ether theory 58

F
False science 328
Fairies 49
Fairy tale 32
Faster-than light speed 184
FBI 318, 320
Fixed star 38, 80, 131
Four laws of relative speed 17, 38, 304
Four real forces 42
4-D coordinate system 22
Four-dimensional unity 5
Four-dimensional space-time 15
Frauenglass 322

G
Galilean reference-body 253
Galilean space 29, 115
Galilean Transformation 173
Galileo 15, 152
Geometry of space-time 45, 258
Geocentricism 38, 49
God 28, 130, 168, 250, 273, 284, 312
GPS 38, 119
Gravitational force 259
Gravitational mass 40, 233
Gravity 120, 253, 259
Gyroscope 14, 148
Gyroscopic inertia 141, 148, 151

H
Harvard physicists 3, 18, 137, 147, 183
Historical sham 263
Hitler 275
Hoax 47, 50
Hypocrite 313

I
Icon 305
Idolatry 328
Imagination 308
Immortality 308
Indictment 3, 18
Inertial force 40, 64, 120, 253
Inertial mass 40, 233
Inertial motion 131, 252
Infant disease 306
Inseparable friends 317
Instantaneous relative speed 20, 26, 59
Instantaneous speed 26, 57
Intermittent acceleration 26
Intersection angle 101
Invisible player 313
Isotropy of light speed 28, 31, 33, 127, 156, 178

J
Jean Foucault 152
Jerk 232, 236, 249, 250
Joy J. Glauber 77

K
Konenkova 320, 321
Korean Physics Society (KPS) 271

L
Law of conservation of matter and energy, the 226
Law of motion 234
Laws of relative motion 74, 304
Law of the equality of gravitational mass and inertial mass, The 39
Leiden (University) 27, 34,

138, 141
Leibniz, Baron von 58
Length contraction 25, 195
Light-year 179
Line 53
Linear speed 61
Linear motion 21, 25, 55, 245, 252
Lorentz factor 24
Lorentz transformation (LT) 33, 173

M

Manifold delusion 5
Marx 275
Mass increase 195
Mathematical delusion 136
Matter point 61, 148, 257
Max Born 313
Maxwell 276
McCarthy 321
Measles 306
Michelson-Morley Experiment (MME) 25, 187
Militant pacifist 317
Misleader 275
Monism 308
Morality 305
Muon paradox 13, 165, 181, 203
Multiverse 17
Murderer 313

N

Naked emperor 328
Nationalism 306
New York Times, The 322
Newton (1642-1727) 152
Newton's bucket 139
Newton's emission theory 27
Newton's law of motion 234
Newton's first law of motion 234
Newton's second law of motion 234, 251
Newton's third law of motion 234
Newton's theory of gravitation 253
Newtonian relativity 30, 43, 240
Non-Euclidian space-time 28
Non-uniform motion 21, 25, 59
Norman Thomas 316, 318
North Star 83
NPA (Natural Philosophy Alliance) 20, 76, 97

O

Observer-free speed 14, 15, 56, 256
Observation angle 90
Observation-angle-priority equation 92
Observation point 21
One-dimensional ideology 16, 89
One-dimensional environment 63
1-D situation 11, 12, 19, 55, 66
One-way experiment 156
One-way light speed 38, 158
Opium of the masses 310
Orbital speed 119
Organized crime 17, 50

P

Pacifist 44, 262
Pantheism 307
Parallel motion 105
Particle accelerator
Paul 275
Perihelion of Mercury 48
Perpendicular 70
Personal God 308
Philosophy 50, 305

Photocentricism 49
Photon 16, 32, 136
Photosphere 265
Plagiarist 47, 48
Plane 53
Poincaré 47
Point 53
Point reference body 21, 63
Polaris 83
Ponderable media 27
Postulate 137, 154
Practical scientists 21
Prayer 309
Principle of equivalence 235
Probability 313
Profound deluder 275
Prominence 266
Proper motion/speed 11, 23, 36, 54, 64, 153
Proper speed of light 29, 33, 56, 127, 128, 137
Proper velocity 11, 87
Proper velocity (relativists' term) 15, 16, 56
Pseudo force 230
Pseudo-gravity 36, 50, 237

Q

Quaker 262, 274
Quantum mechanics 313

R

Railcar 96
Rainbow catching game 195
Rate of actual motion 23
Rate of change in distance 11, 20, 184
Raven 87
Raven thought experiment 87
Reciprocity 11, 31, 64, 129, 133
Rectilinear motion/speed 61, 99, 110, 143, 252
Reference body 21, 63

Refraction 268
Relative speed 11, 55
Relative velocity (relativists' term) 15, 16
Relative velocity 11
Relativistic addition of speeds 163
Relativistic formula of speed addition 12, 33
Relativity of acceleration 60, 231
Relativity of simultaneity 28, 122
Relativity trial 18, 50
Religion 305
r-factor 78, 80, 132, 300, 304
Road to hell 317
Rosenberg 318, 319
Rotation motion 115
Rotatory motion 20, 85
Round-trip average speed of light 158

S

Sagnac 29, 125, 128
Sagnac effect 128
Sagnac experiment 29, 125
Scalar 11, 53
Scientific hoax 47
Sham 263
Single-trip speed of light 158
Slope of space-time 259
Socialist 44
Solar atmosphere 45, 265
Solar eclipse observation 260
Space 53
Spatial distribution of gravitational field 256
Speed 9, 53
Speed detector 20
Speed gun 18, 50, 85, 94, 99
Speedometer 23

Spinoza 307, 309
Steven Weinberg 76
Supernatural Being 309
Super-super government 314
T
Talmud 307
Tangent 12, 78, 99, 113, 148, 303
Tangential speed 61, 114, 120, 196, 256
The first law of relative speed 74, 304
The fourth law of relative speed 83, 150, 300, 303
The law of conservation of matter and energy 226
The law of the equality of gravitational mass and inertial mass 39
The New York Times 322
The second law of relative peed 74, 304
The third law of relative motion 74, 111, 304
Theoretical scientists 21
3-D situation 12, 19, 55, 66
Three perpendicular planes 64
Time dilation 11, 26, 60, 195
Time-factor-priority equation 92
Time travel 17
Top 64, 85
Train thought experiment 28
Translatory motion 20, 85
Twin paradox 12, 129, 164, 199
Two postulates of special relativity 154
Two-dimensional environment 63
2-D situation 12, 19, 55, 66
Two-way experiment 156
Two-way light speed 39
U
Uncertainty principle 313
Unified field theory 276
Uniform acceleration 28, 42, 59, 231, 236, 245
Uniform continuous jerk 236
Uniform motion 21, 25, 59
Universal simultaneity 130.
Universal time 129, 130
Unknown 19, 136
Upper limit of speed 39
V
Variable 19, 136
Vector 11, 53, 281
Velocity 11
Virtual accelerated motion 31
Virtual motion/speed 17, 22, 57, 84, 133, 231
W
Walter Isaacson 305
Weight 40
Wolfgang Pauli 284
World government 307, 314, 315
Worm hole 17
Y
Yuri Bezmenov 321

www.ingramcontent.com/pod-product-compliance
Lightning Source LLC
Chambersburg PA
CBHW071356170526
45165CB00001B/73